REED'S SHIP CON
FOR MARINE S

REED'S SHIP CONSTRUCTION FOR MARINE STUDENTS

E A STOKOE
CEng, FRINA, FIMarE, MNECInst
Formerly Principal Lecturer in Naval Architecture at South Shields
Marine and Technical College

ADLARD COLES NAUTICAL
London

Published by Adlard Coles Nautical
an imprint of Bloomsbury Publishing Plc
50 Bedford Square, London WC1B 3DP
www.adlardcoles.com

First edition published by Thomas Reed Publications 1964
Second edition 1968
Third edition 1973
Reprinted 1975
Fourth edition 1979
Fifth edition 1985
Reprinted 1994, 1996
Reprinted by Adlard Coles Nautical 2005, 2007, 2009, 2010,
2011 (twice) and 2013

ISBN 978-0-7136-7178-0

A CIP catalogue record for this book
is available from the British Library.

This book is produced using paper that is made from wood grown
in managed, sustainable forests. It is natural, renewable and recyclable.
The logging and manufacturing processes conform to the
environmental regulations of the country of origin.

Printed and bound in Great Britain

Note: While all reasonable care has been taken in the publication
of this book, the publisher takes no responsibility for the use of the
methods or products described in the book.

PREFACE

This volume covers the majority of the descriptive work in the Syllabus for Naval Architecture in Part B of the Department of Transport Examinations for Class 2 and Class 1 Engineers together with the ship construction content of the General Engineering Knowledge papers. It is therefore complementary to Volume IV "Naval Architecture for Marine Engineers" and Volume VIII "General Engineering Knowledge" in the same series. It will also be found useful by those studying for Mate and Master's Examinations.

The book is not intended to be comprehensive, but to give an indication of typical methods of construction. The text is concise and profusely illustrated. It is suggested that those engineers studying at sea should first read part of the text, paying particular attention to the diagrams, and then compare the arrangements shown in the book with those on the ship wherever possible. In this way the student will relate the text to the structure. The typical Examination Questions are intended as a revision of the whole work.

The author wishes to acknowledge the considerable assistance given by his former colleagues and to the following firms for permission to use their information and drawings: Fibreglass Ltd, C. M. P. Glands Ltd, Kort Propulsion Co Ltd, Taylor Pallister & Co Ltd, Swan Hunter Shipbuilders Ltd, Welin Davit & Engineering Co Ltd, Brown Bros & Co Ltd, Stone Manganese Marine Ltd, Stone Vickers Ltd and Weir Pumps Ltd.

CONTENTS

CHAPTER 1— **Ship types and terms**   PAGE
Passenger ships, cargo liners, cargo
tramps, oil tankers, bulk carriers,
colliers, container ships, roll-on/roll-
off vessels, liquefied gas carriers,
chemical carriers. Terms in general
use.   1—14

CHAPTER 2— **Stresses in ship structures**
Longitudinal bending in still water and
waves, transverse bending, stresses
when docking, panting and pounding.   15—23

CHAPTER 3— **Sections used: Welding and materials**
Types of rolled steel section used in
shipbuilding. Aluminium sections.
Metallic arc welding, argon arc
welding, types of joint and edge
preparation, advantages and disadvantages, testing of welds, design of
welded structure. Materials, mild steel,
higher tensile steels, Arctic D steel,
aluminium alloys. Brittle fracture.   24—40

CHAPTER 4— **Bottom and side framing**
Double bottom, internal structure,
duct keel, double bottom in machinery
space. Side framing, tank side
brackets, beam knees, web frames.   41—49

CHAPTER 5— **Shell and decks**
Shell plating, bulwarks. Deck plating,
beams, deck girders and pillars, discontinuities, hatches, steel hatch
covers, watertight hatches.   50—60

CHAPTER 6— **Bulkheads and deep tanks**
Watertight bulkheads, watertight
doors. Deep tanks for water ballast

and for oil. Non-watertight bulkheads, corrugated bulkheads.   61—72

CHAPTER 7— **Fore end arrangements**
Stem plating, arrangements to resist panting, arrangements to resist pounding, bulbous bow, anchor and cable arrangements.   73—81

CHAPTER 8— **After end arrangements**
Cruiser stern, transom stern, sternframe and rudder, fabricated sternframe, cast steel sternframe, unbalanced rudder, balanced rudder, open water stern, spade rudder, rudder and sternframe for twin screw ship. Bossings and spectacle frame. Shaft tunnel. Kort nozzle, fixed nozzle, nozzle rudder. Tail flaps and rotary cylinders.   82—98

CHAPTER 9— **Oil tankers, bulk carriers, liquefied gas carriers and container ships**
Oil tankers, longitudinal framing, combined framing, cargo pumping and piping, crude oil washing. Bulk carriers, ore carriers. Liquefied gas carriers, fully pressurised, semi-pressurised/partly refrigerated, semi-pressurised/fully refrigerated, fully refrigerated, safety and environmental control, boil off, operating procedures. Container ships.   99—122

CHAPTER 10— **Freeboard, tonnage, life saving appliances, fire protection and classification**
Freeboard definitions, basis for calculation, markings, conditions of assignment, surveys. Tonnage, 1967 rules, definitions, underdeck, gross and net tonnage, propelling power allowance, modified tonnage, alternative tonnage. 1982 rules, gross

and net tonnage calculation. Life
saving appliances, lifeboats, davits.
Fire protection, definitions, passenger
ships, dry cargo ships, oil tankers.
Classification of ships, assignment of
class, surveys, discontinuities. 123—140

CHAPTER 11— **Ship Dynamics**
Propellers, wake distribution, blade
loading, controllable pitch propellers,
contra-rotating propellers, vertical axis
propellers. Bow thrusters, controllable
pitch thrusters, hydraulic thrust units.
Rolling and stabilisation, reduction of
roll, bilge keels, fin stabilisers, tank
stabilisers. Vibration, causes and
reduction. 141-158

CHAPTER 12— **Miscellaneous**
Insulation of ships. Corrosion, prevention, surface preparation, painting,
cathodic protection, impressed current
system, design and maintenance.
Fouling. Examination in dry dock.
Emergency repairs to structure. Engine
casing. Funnel. 159—175

**Selection of Examination Questions —
Class 2**   176—180

**Selection of Examination Questions —
Class 1**   181—187

Index   188—192

CHAPTER 1

SHIP TYPES AND TERMS

Merchant ships vary considerably in size, type, layout and function. They include passenger ships, cargo ships and specialised types suitable for particular classes of work. This book deals with the construction of normal types of passenger ships and cargo ships. The cargo ships may be subdivided into those designed to carry various cargoes and those intended to carry specific cargoes, such as oil tankers, bulk carriers and colliers.

PASSENGER SHIPS

A *passenger ship* may be defined as one which may accommodate more than 12 passengers. They range from small river ferries to large ocean-going vessels which are in the form of floating hotels. The larger ships are designed for maximum comfort to large numbers of passengers, and include in their services large dining rooms, lounges suitable for dances, cinemas, swimming pools, gymnasia, open deck spaces and shops. They usually cater for two or three classes of passenger, from tourist class to the more luxurious first class. Where only a small number of passengers is carried in comparison with the size of the ship, the amenities are reduced.

Any ships travelling between definite ports and having particular departure and arrival dates are termed *liners*. Thus a passenger liner is one which travels between particular ports. Because of their rigid timetable such ships are often used for carrying mail and perishable goods in their greatly restricted cargo space, their high speeds ensuring minimum time on passage.

The regulations enforced for the construction and maintenance of passenger ships are much more stringent than those for cargo ships in an attempt to provide safe sea passage.

Many of the regulations are the result of losses of ships which were previously regarded as safe, sometimes with appalling loss of life.

CARGO LINERS

Cargo liners are vessels designed to carry a variety of cargoes between specific ports. It is usual in these ships to carry a cargo of a 'general' nature, *i.e.,* an accumulation of smaller loads from different sources, although many have refrigerated compartments capable of carrying perishable cargoes such as meat, fruit and fish. These vessels are termed reefers. Arrangements are often made to carry up to 12 passengers. These ships are designed to run at speeds of between 15 knots and 20 knots.

Fig. 1.1 shows the layout of a modern, two-deck cargo liner. At the extreme fore end is a tank known as the *fore peak* which may be used to carry water ballast or fresh water. Above this tank is a *chain locker* and *store space*. At the after end is a tank known as the *after peak* enclosing the stern tube in a watertight compartment. Between the peak bulkheads is a continuous tank top forming a *double bottom* space which is subdivided into tanks suitable for carrying oil fuel, fresh water and water ballast. The *machinery space* is shown aft of midships presenting an uneven distribution of cargo space. This is a modern arrangement and slightly unusual, but has the effect of reducing the maximum bending moment. A more usual design in existing ships has the machinery space near midships, with three holds forward and two aft, similar to the arrangement shown in Fig. 1.2. The *oil fuel bunkers* and *settling tanks* are arranged adjacent to, or at the side of, the machinery space. From the after engine room bulkhead to the after peak bulkhead is a watertight *shaft tunnel* enclosing the shaft and allowing access to the shaft and bearings directly from the engine room. An exit in the form of a vertical trunk is arranged at the after end of the tunnel in case of emergency. In a twin screw ship it is necessary to construct two such tunnels, although they may be joined together at the fore and after ends.

The cargo space is divided into lower *holds* and compartments between the decks, or *'tween decks*. Many ships have three decks, thus forming upper and lower 'tween decks. This system allows different cargoes to be carried in different compartments and reduces the possibility of crushing the cargo. Access to the cargo compartments is provided by means of large hatchways

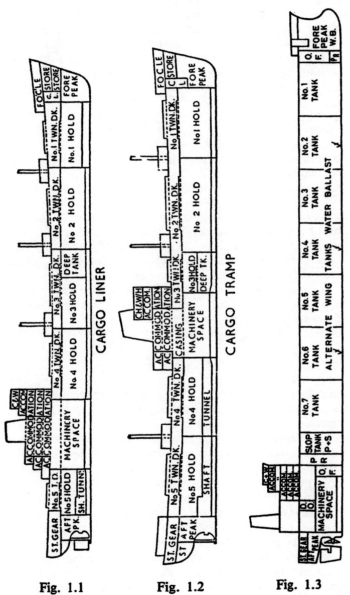

Fig. 1.1   Fig. 1.2   Fig. 1.3

which may be closed either by wood boards or by steel covers, the latter being most popular in modern ships. Suitable cargo handling equipment is provided in the form of either derricks or cranes. Heavy lift equipment is usually fitted in way of one hatch. A *forecastle* is fitted to reduce the amount of water shipped forward and to provide adequate working space for handling ropes and cables.

CARGO TRAMPS

Cargo tramps are those ships which are designed to carry no specific type of cargo and travel anywhere in the world. They are often run on charter to carry bulk cargo or general cargo, and are somewhat slower than the cargo liners. Much of their work is being taken over by bulk carriers.

Fig. 1.2 shows the layout of a typical cargo tramp. The arrangement of this ship is similar to that shown for the cargo liner, except that the machinery space is amidships. The space immediately forward of the machinery space is subdivided into a lower 'tween decks and *hold/deep tank*. Many ships have no such subdivision, the compartment being alternatively a hold or a deep tank depending upon whether the ship carries cargo or is in a ballast condition. The former arrangement has the advantage of reducing the stresses in the ship if, in the loaded condition, the deep tank is left empty.

OIL TANKERS

Tankers are used to carry oil or other liquids in bulk, oil being the most usual cargo. The machinery is situated aft to provide an unbroken cargo space which is divided into tanks by longitudinal and transverse bulkheads. The tanks are separated from the machinery space by an empty compartment known as a cofferdam. A pump room is provided at the after end of the cargo space and may form part of the cofferdam. (Fig. 1.3).

A double bottom is required only in way of the machinery space and may be used for the carriage of oil fuel and fresh water. A forecastle is sometimes required and is used as a store space. The accommodation and navigation spaces are provided at the after end, leaving the deck space unbroken by super-structure and concentrating all the services and catering equipment in one area. Much of the deck space is taken by pipes and hatches. It is usual to provide a longitudinal platform to allow easy access to the fore end, above the pipes.

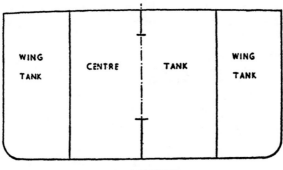

OIL TANKER

Fig. 1.4

The midship section (Fig. 1.4) shows the transverse arrangement of the cargo tanks. The centre tank is usually about half the width of the ship.

BULK CARRIERS

Bulk carriers are vessels built to carry such cargoes as ore, coal, grain and sugar in large quantities. They are designed for ease of loading and discharging with the machinery space aft, allowing continuous, unbroken cargo space. They are single deck vessels having long, wide hatches, closed by steel covers. The double bottom runs from stem to stern. In ships designed for heavy cargoes such as iron ore the double bottom is very deep and longitudinal bulkheads are fitted to restrict the cargo space (Fig. 1.5). This system raises the centre of gravity of the ore, resulting in a more comfortable ship. The double bottom and the wing compartments may be used as ballast tanks for the return voyage. Some vessels, however, are designed to carry an alternative cargo of oil in these tanks. With lighter cargoes such as grain, the restriction of the cargo spaces is not necessary although deep hopper sides are fitted to facilitate the discharge of cargo, either by suction or grabs. The spaces at the sides of the hatches are plated in as shown in Fig. 1.6 to give self trimming properties. In many bulk carriers a tunnel is fitted below the deck from the midship superstructure to the accommodation at the after end. The remainder of the wing space may be used for water ballast. Some bulk carriers are built with alternate long and short compartments. Thus if a heavy cargo such as iron ore is carried, it is loaded into the short holds.

A cargo such as bauxite would be carried in the long holds, whilst a light cargo such as grain or timber would occupy the whole hold space.

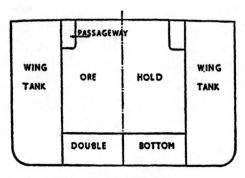

ORE CARRIER

Fig. 1.5

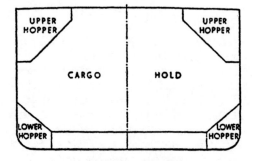

BULK CARRIER

Fig. 1.6

COLLIERS

Colliers are usually much smaller than the usual range of bulk carriers, being used mainly for coastal trading. Fig. 1.8 shows the layout of a modern collier.

The machinery space is again aft, but in small vessels this creates a particular problem. The machinery itself is heavy, but the volume of the machinery space is relatively large. Thus the weight of the machinery is much less than the weight of a normal cargo which could be carried in the space. In the lightship or ballast condition, the ship trims heavily by the stern, but in the loaded condition the weight of cargo forward would normally exceed the weight of machinery aft, causing the vessel to trim by the head. It is usual practice in colliers, and in most other coastal

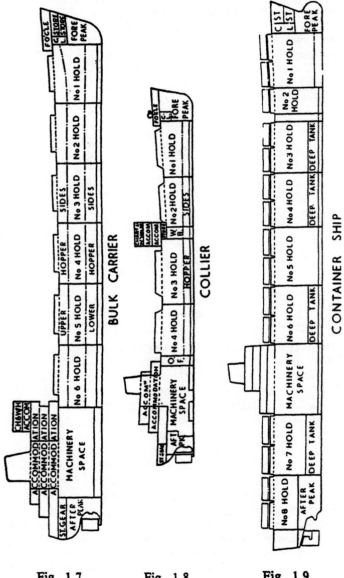

Fig. 1.7   Fig. 1.8   Fig. 1.9

vessels, to raise the level of the upper deck aft, providing a greater volume of cargo space aft. This forms a *raised quarter deck ship*.

The double bottom is continuous in the cargo space, being knuckled up at the bilges to form hopper sides which improve the rate of discharge of cargo. In way of the machinery space a double bottom is fitted only in way of the main machinery, the remainder of the space having open floors. Wide hatches are fitted for ease of loading, while in some ships small wing tanks are fitted to give self trimming properties. Fig. 1.10 shows the transverse arrangement of the cargo space.

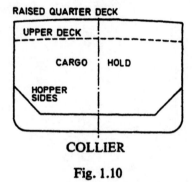

COLLIER

Fig. 1.10

CONTAINER SHIPS

The cost of cargo handling in a general cargo ship is about 40% of the total running costs of the ship. An attempt has been made to reduce these costs by reducing the number of items lifted, *i.e.*, by using large rectangular containers. These containers are packed at the factory and opened at the final delivery point, thus there is less chance of damage and pilfering. They are fitted with lifting lugs to reduce transfer time.

Most efficient use is made of such containers when the whole transport system is designed for this type of traffic, *i.e.*, railway trucks, lorries, lifting facilities, ports and ships. For this reason fast container ships have been designed to allow speedy transfer and efficient stowage of containers. These vessels have rectangular holds thus reducing the cargo capacity, but this is more than compensated by the reduced cargo handling costs and increased speed of discharge. Fig. 1.9 shows a typical arrangement of a container ship with containers stowed above the deck.

ROLL-ON/ROLL-OFF VESSELS

In order to reduce the cargo handling costs and time in port further, vessels have been designed with flat decks which are virtually unrestricted. A ramp is fitted at the after end allowing direct access to cars, trucks and trailers which remain on board in their laden state. Similarly containers may be loaded two or three high by means of fork lift trucks. Lifts and inter-deck ramps are used to transfer vehicles between decks. Modern ramps are angled to allow vehicles to be loaded from a straight quay. Accommodation is provided for the drivers and usually there is additional passenger space since Ro-Ro's tend to work as liners.

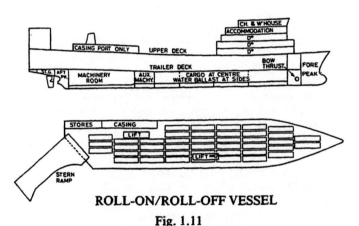

ROLL-ON/ROLL-OFF VESSEL

Fig. 1.11

LIQUEFIED GAS CARRIERS

The discovery of large reservoirs of natural gas has led to the building of vessels equipped to carry the gas in liquefied form. The majority of gas carried in this way is methane which may be liquefied by reducing the temperature to between $-82°C$ and $-162°C$ in association with pressures of 4.6 MN/m^2 to atmospheric pressure. Since low carbon steel becomes extremely brittle at low temperatures, separate containers must be built within the hull and insulated from the hull. Several different systems are available, one of which is shown in Fig. 1.12 and Fig. 1.14. The cargo space consists of three large tanks set in about 1 m from the ship's side. Access is provided around the sides and ends of the tanks, allowing the internal structure to be inspected.

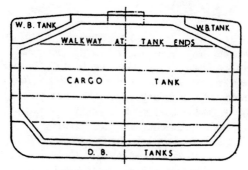

LIQUEFIED GAS CARRIER

Fig. 1.12

CHEMICAL CARRIERS

A considerable variety of chemical cargoes are now required to be carried in bulk. Many of these cargoes are highly corrosive and incompatible while others require close control of temperature and pressure. Special chemical carriers have been designed and built, in which safety and avoidance of contamination are of prime importance.

To avoid corrosion of the structure, stainless steel is used extensively for the tanks, while in some cases coatings of zinc silicate or polyurethane are acceptable.

Protection for the tanks is provided by double bottom tanks and wing compartments which are usually about one fifth of the midship beam from the ship side. (Fig. 1.13).

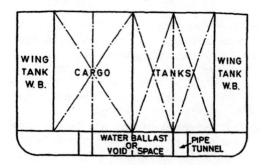

CHEMICAL CARRIER

Fig. 1.13

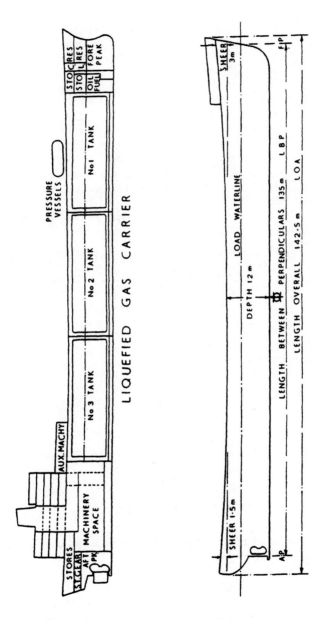

Fig. 1.14

Fig. 1.15

GENERAL NOTES

Ocean-going ships must exist as independent units. Cargo handling equipment suitable for the ship's service is provided. Navigational and radio equipment of high standard is essential. The main and auxiliary machinery must be sufficient to propel the ship at the required speed and to maintain the ship's services efficiently and economically. Adequate accommodation is provided for officers and crew with comfortable cabins, recreation rooms and dining rooms. Air conditioning or mechanical ventilation is fitted because of the tremendous variation in air temperature. Many ships have small swimming pools of the portable or permanent variety. The ships must carry sufficient foodstuffs in refrigerated and non-refrigerated stores for the whole trip, together with ample drinking water. In the event of emergency it is essential that first aid, fire extinguishing and life saving appliances are provided.

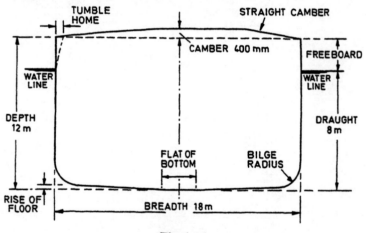

Fig. 1.16

SHIP TERMS

The following terms and abbreviations are in use throughout the shipbuilding industry.

Length overall (L.O.A.)
The distance from the extreme fore part of the ship to a similar point aft and is the greatest length of the ship. This length is important when docking.

Length between perpendiculars (L.B.P.)

The fore perpendicular is the point at which the Summer Load Waterline crosses the stem. The after perpendicular is the after side of the rudder post or the centre of the rudder stock if there is no rudder post. The distance between these two points is known as the length between perpendiculars, and is used for ship calculations.

Breadth extreme (B. Ext)

The greatest breadth of the ship, measured to the outside of the shell plating.

Breadth moulded (B. Mld)

The greatest breadth of the ship, measured to the inside of the inside strakes of shell plating.

Depth extreme (D. Ext)

The depth of the ship measured from the underside of the keel to the top of the deck beam at the side of the uppermost continuous deck amidships.

Depth moulded (D. Mld)

The depth measured from the top of the keel.

Draught extreme (d. Ext)

The distance from the bottom of the keel to the waterline. The load draught is the maximum draught to which a vessel may be loaded.

Draught moulded (d. Mld)

The draught measured from the top of the keel to the waterline.

Freeboard

The distance from the waterline to the top of the deck plating at the side of the deck amidships.

Camber or round of beam

The transverse curvature of the deck from the centreline down to the sides. This camber is used on exposed decks to drive water to the sides of the ship. Other decks are often cambered. Most modern ships have decks which are flat transversely over the width of the hatch or centre tanks and slope down towards the side of the ship.

Sheer

The curvature of the deck in a fore and aft direction, rising from midships to a maximum at the ends. The sheer forward is usually twice that aft. Sheer on exposed decks makes a ship more seaworthy by raising the deck at the fore and after ends further from the water and by reducing the volume of water coming on the deck.

Rise of floor

The bottom shell of a ship is sometimes sloped up from the keel to the bilge to facilitate drainage. This rise of floor is small, 150 mm being usual.

Bilge radius

The radius of the arc connecting the side of the ship to the bottom at the midship portion of the ship.

Tumble home

In some ships the midship side shell in the region of the upper deck is curved slightly towards the centreline, thus reducing the width of the upper deck and decks above. Such tumble home improves the appearance of the ship.

Displacement

The mass of the ship and everything it contains. A ship has different values of displacement at different draughts.

Lightweight

The mass of the empty ship, without stores, fuel, water, crew or their effects.

Deadweight

The mass of cargo, fuel, water, stores, etc., a ship carries. The deadweight is the difference between the displacement and the lightweight

i.e., displacement = lightweight + deadweight

It is usual to discuss the size of a cargo ship in relation to its deadweight. Thus a 10 000 tonne ship is one which is capable of carrying a deadweight of 10 000 tonne.

The dimensions given in Figs. 1.15 and 1.16 are typical for a ship of about 10 000 tonne deadweight.

Liquefied gas carriers are compared in terms of the capacity of the cargo tanks, e.g., 10 000 m³.

CHAPTER 2

STRESSES IN SHIP STRUCTURES

Numerous forces act on a ship's structure, some of a static nature and some dynamic. The static forces are due to the differences in weight and support which occur throughout the ship, while the dynamic forces are created by the hammering of the water on the ship, the passage of waves along the ship and by the moving machinery parts. The greatest stresses set up in the ship as a whole are due to the distribution of loads along the ship, causing longitudinal bending.

LONGITUDINAL BENDING

A ship may be regarded as non-uniform beam, carrying non-uniformly distributed weights and having varying degrees of support along its length.

(a) Still water bending

Consider a loaded ship lying in still water. The upthrust at any one metre length of the ship depends upon the immersed cross-sectional area of the ship at that point. If the values of upthrust at different positions along the length of the ship are plotted on a base representing the ship's length, a *buoyancy curve* is formed (Fig. 2.1). This curve increases from zero at each end to a maximum value in way of the parallel midship portion. The area of this curve represents the total upthrust exerted by the water on the ship. The total weight of a ship consists of a number of independent weights concentrated over short lengths of the ship, such as cargo, machinery, accommodation, cargo handling gear, poop and forecastle, and a number of items which form continuous material over the length of the ship, such as decks, shell and tank top. A *curve of weights* is shown in Fig. 2.1. The difference between the weight and buoyancy at any point is the load at that point. In some cases the load is an excess of weight over buoyancy and in other cases an excess of buoyancy over weight. A *load diagram* formed by these differences is shown in the figure. Since the total weight must be equal to the total buoyancy, the area of the load diagram above the base line must be equal to the area below the base line. Because of this unequal loading, however, shearing forces and bending moments are set up in the ship. The maximum bending moment occurs about midships.

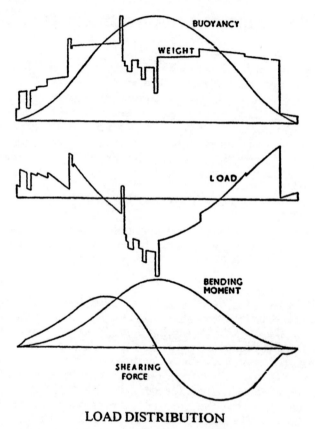

LOAD DISTRIBUTION

Fig. 2.1

Depending upon the direction in which the bending moment acts, the ship will hog or sag. If the buoyancy amidships exceeds the weight, the ship will hog, and may be likened to a beam supported at the centre and loaded at the ends.

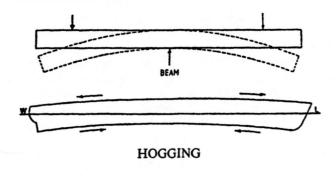

HOGGING

Fig. 2.2

When a ship hogs, the deck structure is in tension while the bottom plating is in compression (Fig. 2.2).

If the weight amidships exceeds the buoyancy, the ship will sag, and is equivalent to a beam supported at its ends and loaded at the centre.

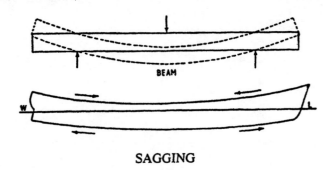

SAGGING

Fig. 2.3

When a ship sags, the bottom shell is in tension while the deck is in compression (Fig. 2.3).

Changes in bending moment occur in a ship due to different systems of loading. This is particularly true in the case of cargoes such as iron ore which are heavy compared with the volume they occupy. If such cargo is loaded in a tramp ship, care must be taken to ensure a suitable distribution throughout the ship. Much trouble has been found in ships having machinery space and deep tank/cargo hold amidships. There is a tendency in such ships, when loading heavy cargoes, to leave the deep tank empty. This results in an excess of buoyancy in way of the deep tank. Unfortunately there is also an excess of buoyancy in way of the engine room, since the machinery is light when compared with the volume it occupies. A ship in such a loaded condition would therefore hog, creating very high stresses in the deck and bottom shell. This may be so dangerous that if owners intend the ships to be loaded in this manner, additional deck material must be provided.

The structure resisting longitudinal bending consists of all continuous longitudinal material, the portions farthest from the axis of bending (the neutral axis) being the most important (Fig. 2.4), *e.g., keel,* bottom shell, *centre girder,* side girders, tank top, tank margin, side shell, *sheerstrake, stringer plate,* deck plating alongside hatches, and in the case of oil tankers, longitudinal bulkheads. Danger may occur where a *point* in the structure is the greatest distance from the neutral axis, such as the top of a sheerstrake, where a high stress point occurs. Such points are to be avoided as far as possible, since a crack in the plate may result. In many oil tankers the structure is improved by joining the sheerstrake and stringer plate to form a rounded gunwale.

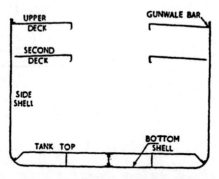

LONGITUDINAL MATERIAL

Fig. 2.4

(b) **Wave bending**

When a ship passes through waves, alterations in the distribution of buoyancy cause alterations in the bending moment. The greatest differences occur when a ship passes through waves whose lengths from crest to crest are equal to the length of the ship.

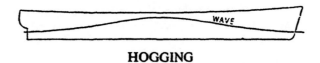

HOGGING

Fig. 2.5

When the wave crest is amidships (Fig. 2.5), the buoyancy amidships is increased while at the ends it is reduced. This tends to cause the ship to hog.

SAGGING

Fig. 2.6

A few seconds later the wave trough lies amidships. The buoyancy amidships is reduced while at the ends it is increased, causing the vessel to sag (Fig. 2.6).

The effect of these waves is to cause fluctuations in stress, or, in extreme cases, complete reversals of stress every few seconds. Fortunately such reversals are not sufficiently numerous to cause fatigue, but will cause damage to any faulty part of the structure.

TRANSVERSE BENDING

The transverse structure of a ship is subject to three different types of loading:

(a) forces due to the weights of the ship structure, machinery, fuel, water and cargo.

(b) water pressure.

(c) forces created by longitudinal bending.

The decks must be designed to support the weight of accommodation, winches and cargo, while exposed decks may have to withstand a tremendous weight of water shipped in heavy weather. The deck plating is connected to beams which transmit the loads to longitudinal girders and to the side frames. In way of heavy local loads such as winches, additional stiffening is arranged. The shell plating and frames form pillars which support the weights from the decks. The tank top is required to carry the weight of the hold cargo or the upthrust exerted by the liquid in the tanks, the latter usually proving to be the most severe load.

In the machinery space other factors must be taken into account. Forces of pulsating nature are transmitted through the structure due to the general out of balance forces of the machinery parts. The machinery seats must be extremely well supported to prevent any movement of the machinery. Additional girders are fitted in the double bottom and the thickness of the tank top increased under the engine in an attempt to reduce the possibility of movement which could cause severe vibration in the ship. For similar reasons the shaft and propeller must be well supported.

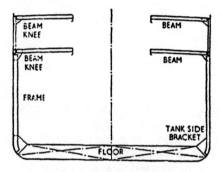

TRANSVERSE MATERIAL

Fig. 2.7

A considerable force is exerted on the bottom and side shell by the water surrounding the ship. The double bottom floors and side frames are designed to withstand these forces, while the shell plating must be thick enough to prevent buckling between the floors and frames. Since water pressure increases with the depth of immersion, the load on the bottom shell exceeds that on the side shell. It follows, therefore, that the bottom shell must be thicker than the side shell. When the ship passes through waves, these forces are of a pulsating nature and may vary considerably in high waves, while in bad weather conditions the shell plating above the waterline will receive severe hammering.

When a ship rolls there is a tendency for the ship to distort transversely in a similar way to that in which a picture frame may collapse. This is known as *racking* and is reduced or prevented by the beam knee and tank side bracket connections, together with the transverse bulkheads, the latter having the greatest effect.

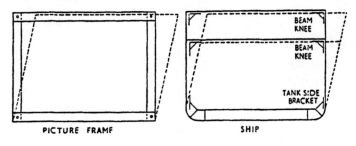

RACKING

Fig. 2.8

The efficiency of the ship structure in withstanding longitudinal bending depends to a large extent on the ability of the transverse structure to prevent collapse of the shell plating and decks.

Docking

A ship usually enters dry dock with a slight trim aft. Thus as the water is pumped out, the after end touches the blocks. As more water is pumped out an upthrust is exerted by the blocks on the after end, causing the ship to change trim until the whole keel from forward to aft rests on the centre blocks. At the instant before this occurs the upthrust aft is a maximum. If this thrust is excessive it may be necessary to strengthen the after blocks and the after end of the ship. Such a problem arises if it is necessary to dock a ship when fully loaded or when trimming severely by the stern. As the pumping continues the load on the keel blocks is increased until the whole weight of the ship is taken by them. The ship structure in way of the keel must be strong enough to withstand this load. In most ships the normal arrangement of keel and centre girder, together with the transverse floors, is quite sufficient for the purpose. If a duct keel is fitted, however, care must be taken to ensure that the width of the duct does not exceed the width of the keel blocks. The keel structure of an oil tanker is strengthened by fitting docking brackets, tying the centre girder to the adjacent longitudinal frames at intervals of about 1.5 m.

Bilge blocks or shores are fitted to support the sides of the ship. The arrangements of the bilge blocks vary from dock to

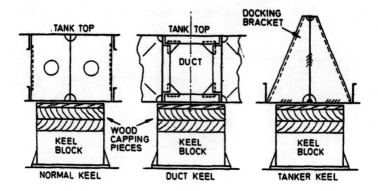

Fig. 2.9

dock. In some cases they are fitted after the water is out of the dock, while some docks have blocks which may be slid into place while the water is still in the dock. The latter arrangement is preferable since the sides are completely supported. At the ends of the ship, the curvature of the shell does not permit blocks to be fitted and so bilge shores are used. The structure at the bilge must prevent these shores and blocks buckling the shell.

As soon as the after end touches the blocks, shores are inserted between the stern and the dock side, to centralise the ship in the dock and to prevent the ship slipping off the blocks. When the ship grounds along its whole length additional shores

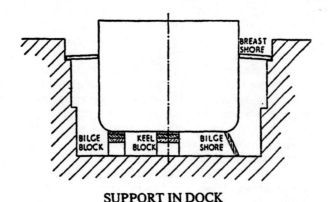

SUPPORT IN DOCK

Fig. 2.10

are fitted on both sides, holding the ship in position and preventing tipping. These shores are known as *breast shores* and have some slight effect in preventing the side shell bulging. They should preferably be placed in way of transverse bulkheads or side frames.

Pounding

When a ship meets heavy weather and commences heaving and pitching, the rise of the fore end of the ship occasionally synchronises with the trough of a wave. The fore end then emerges from the water and re-enters with a tremendous slamming effect, known as pounding. While this does not occur with great regularity, it may nevertheless cause damage to the bottom of the ship forward. The shell plating must be stiffened to prevent buckling. Pounding also occurs aft in way of the cruiser stern but the effects are not nearly as great.

Panting

As the waves pass along the ship they cause fluctuations in water pressure which tend to create an in-and-out movement of the shell plating. The effect of this is found to be greatest at the ends of the ship, particularly at the fore end, where the shell is relatively flat. Such movements are termed panting and, if unrestricted, could eventually lead to fatigue of the material and must therefore be prevented.

The structure at the ends of the ship is stiffened to prevent any undue movement of the shell.

CHAPTER 3

SECTIONS USED:
WELDING AND MATERIALS

When iron was used in the construction of ships in preference to wood, it was found necessary to produce forms of the material suitable for connecting plates and acting as stiffeners. These forms were termed sections and were produced by passing the material through suitably shaped rolls. The development of these bars continued with the introduction of steel until many different sections were produced. These sections are used in the building of modern ships and are known as *rolled steel sections*.

Ordinary angles

These sections may be used to join together two plates meeting at right angles or to form light stiffeners in riveted ships. Two types are employed, those having equal flanges (Fig. 3.1), varying in size between 75 mm and 175 mm, and those having unequal flanges (Fig. 3.2), which may be obtained in a number of sizes up to 250 mm by 100 mm, the latter type being used primarily as stiffeners.

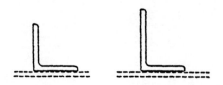

EQUAL ANGLE   UNEQUAL ANGLE

Fig. 3.1   Fig. 3.2

In welded ships, connecting angles are no longer required but use may be made of the unequal angles by toe-welding them to the plates, forming much more efficient stiffeners (Fig. 3.3).

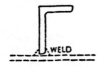

TOE WELDED ANGLE

Fig. 3.3

Bulb angles
The bulb at the toe of the web increases the strength of the bar considerably, thus forming a very economical stiffening member in riveted ships (Fig. 3.4). Bulb angles vary in depth between 115 mm and 380 mm and are used throughout the ship for frames, beams, bulkhead stiffeners and hatch stiffeners.

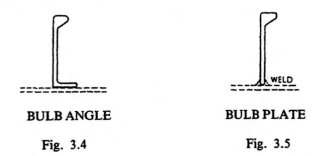

BULB ANGLE   BULB PLATE

Fig. 3.4   Fig. 3.5

Bulb plates
In welded construction the flange of the bulb angles is superfluous, increasing the weight of the structure without any appreciable increase in strength, since it is not required for connection purposes. A bulb plate (Fig. 3.5) has therefore been especially developed for welded construction, having a bulb slightly heavier than the equivalent bulb angle. A plate having a bulb on both sides has been available for many years but its use has been severely limited due to the difficulty of attaching brackets to the web in way of the bulb. The modern section resolves this problem since the brackets may be either overlapped or butt welded to the flat portion of the bulb. Such

sections are available in depths varying between 80 mm and 430 mm, being lighter than the bulb angles for equal strength. They are used for general stiffening purposes in the same way as bulb angles.

Channels

Channel bars (Fig. 3.6) are available in depths varying between 160 mm and 400 mm. Channels are used for panting beams, struts, pillars and girders and heavy frames. In insulated ships it is necessary to provide the required strength of bulkheads, decks and shell with a minimum depth of stiffener and at the same time provide a flat inner surface for connecting the facing material in order to reduce the depth of insulation required and to provide maximum cargo space. In many cases, therefore, channel bars with reverse bars are used for such stiffening (Fig. 3.7), reducing the depth of the members by 50 mm or 75 mm. Both the weight and the cost of this method of construction are high.

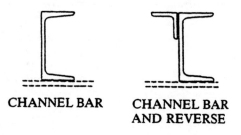

CHANNEL BAR   CHANNEL BAR
AND REVERSE

Fig. 3.6   Fig. 3.7

Joist or H-bars

These sections have been used for many years for such items as crane rails but have relatively small flanges. The manufacturers have now produced such sections with wide flanges (Fig. 3.8), which prove much more useful in ship construction. They are used for crane rails, struts and pillars, being relatively strong in all directions. In deep tanks and engine rooms where tubular pillars are of little practical use, the broad flanged beam may be used to advantage.

Tee bars

The use of the T-bar (Fig. 3.9) is limited in modern ships. Occasionally they are toe-welded to bulkheads (Fig. 3.10) to form heavy stiffening of small depth. Many ships have bilge keels incorporating T-bars in the connection to the shell.

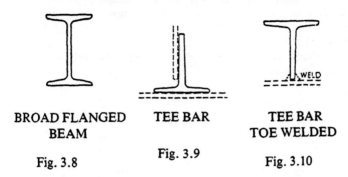

BROAD FLANGED BEAM

Fig. 3.8

TEE BAR

Fig. 3.9

TEE BAR TOE WELDED

Fig. 3.10

Flat bars or slabs

Flat bars are often used in ships of welded construction, particularly for light stiffening, waterways, and save-alls which prevent the spread of oil. Large flat bars are used in oil tankers and bulk carriers for longitudinal stiffening where the material tends to be in tension or compression rather than subject to high bending moments. This allows for greater continuity in the vicinity of watertight or oiltight bulkheads.

Several other sections are used in ships for various reasons. Solid round bars (Fig. 3.11) are used for light pillars, particularly in accommodation spaces, for welded stems and for fabricated rudders and stern frames. Half-round bars (Fig. 3.12) are used for stiffening in accommodation where projections may prove dangerous (*e.g.,* in toilets and wash places), and for protection of ropes from chafing.

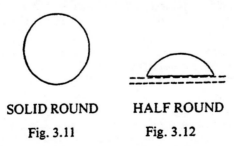

SOLID ROUND

Fig. 3.11

HALF ROUND

Fig. 3.12

Aluminium sections

Aluminium alloys used in ship construction are found to be too soft to roll successfully in section form, and are therefore produced by extrusion, *i.e.*, forcing the metal through a suitably shaped die. This becomes an advantage since the dies are relatively cheap to produce, allowing numerous shapes of section to be made. Thus there are few *standard* sections but the aluminium companies are prepared to extrude any feasible forms of section which the shipbuilders require in reasonable quantities. Fig. 3.13 shows some such sections which have been produced for use on ships built in this country.

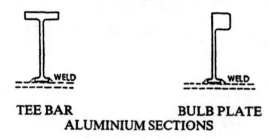

TEE BAR   BULB PLATE
ALUMINIUM SECTIONS

Fig. 3.13

WELDING

Welding is the science of joining two members together in such a way that they become one integral unit. It exists in many different forms from the forging carried out by blacksmiths to the modern electric welding. There are two basic types of welding, resistance or pressure welding in which the portions of metal are brought to a welding temperature and an applied force is used to form the joint, and fusion welding where the two parts forming the joint are raised to a melting temperature and either drawn together or joined by means of a filler wire of the same material as the adjacent members. The application of welding to shipbuilding is almost entirely restricted to fusion welding in the form of metallic arc welding.

Metallic arc welding.
Fig. 3.14 shows a simplified circuit used in arc welding.
A metal electrode, of the same material as the workpiece, is clamped into a holder which is connected to one terminal of a welding unit, the opposing terminal being connected to the workpiece. An arc is formed between the electrode and the workpiece in way of the joint, creating an extreme temperature

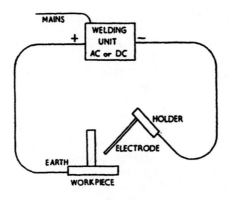

WELDING CIRCUIT

Fig. 3.14

which melts the two parts of the joint and the electrode. Metal particles from the electrode then bombard the workpiece, forming the weld. The arc and the molten metal must be protected to prevent oxidation. In the welding of steel a coated electrode is used, the coating being in the form of a silicone. This coating melts at a slightly slower rate than the metal and is carried with the particles to form a slag over the molten metal, while at the same time an inert gas is formed which shields the arc (Fig. 3.15).

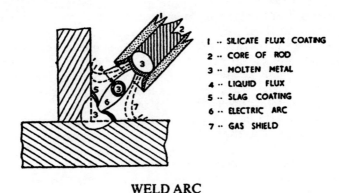

1 .. SILICATE FLUX COATING
2 .. CORE OF ROD
3 .. MOLTEN METAL
4 .. LIQUID FLUX
5 .. SLAG COATING
6 .. ELECTRIC ARC
7 .. GAS SHIELD

WELD ARC

Fig. 3.15

The slag must be readily removed by chipping when cooled.

Argon arc welding

It is found that with some metals, such as aluminium, coated electrodes may not be used. The coatings cause the aluminium to corrode and, being heavier than the aluminium, remain trapped in the weld. It is nevertheless necessary to protect the arc and an inert gas such as argon may be used for this purpose. In argon arc welding, argon is passed through a tube, down the centre of which is a tungsten electrode. An arc is formed between the workpiece and the electrode while the argon forms a shield around the arc. A separate filler wire of suitable material is used to form the joint. The tungsten electrode must be water cooled. This system of welding may be used for most metals and alloys, although care must be taken when welding aluminium to use a.c. supply.

Types of joint and edge preparation

The most efficient method of joining two plates which lie in the same plane is by means of a butt weld, since the two plates then become one continuous member. A square-edge butt (Fig. 3.16) may be used for plates up to about 10 mm thick. Above this thickness, however, it is difficult to obtain sufficient penetration and it becomes necessary to use single vee (Fig. 3.17) or double vee butts (Fig. 3.18). The latter are more economical as far as the volume of weld metal is concerned, but may require more overhead welding and are therefore used only for large thicknesses of plating. The edge preparations for all these joints may be obtained by means of profile burners having three burning heads which may be adjusted to suit the required angle of the joint (Fig. 3.19).

Overlap joints (Fig. 3.20) may be used in place of butt welds, but are not as efficient since they do not allow complete penetration of the material and transmit a bending moment to the weld metal. Such joints are used in practice, particularly when connecting brackets to adjacent members.

Fillet welds (Fig. 3.21) are used when two members meet at right angles. The strength of these welds depends upon the leg length and the throat thickness, the latter being at least 70% of the leg length. The welds may be continuous on one or both sides of the member or may be intermittent. Continuous welds are used when the joint must be watertight and for other strength members.

Stiffeners, frames and beams may be connected to the plating by intermittent welding (Fig. 3.22). In tanks, however, where the rate of corrosion is high, such joints may not be used and it is

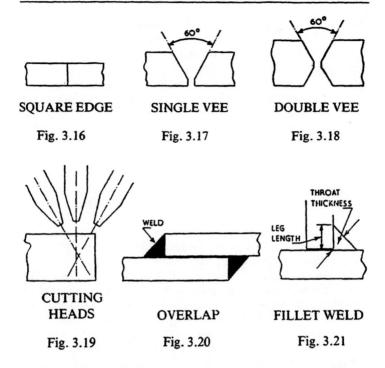

SQUARE EDGE
Fig. 3.16

SINGLE VEE
Fig. 3.17

DOUBLE VEE
Fig. 3.18

CUTTING
HEADS
Fig. 3.19

OVERLAP
Fig. 3.20

FILLET WELD
Fig. 3.21

then necessary to employ continuous welding or to scallop the
section (Fig. 3.23). The latter method has the advantage of
reducing the weight of the structure and improving the drainage,
although a combination of corrosion and erosion may reduce
the section between the scallops.

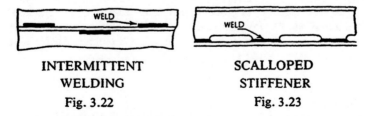

INTERMITTENT
WELDING
Fig. 3.22

SCALLOPED
STIFFENER
Fig. 3.23

Advantages and disadvantages

The amount of welding employed in shipbuilding since the
end of World War II has increased tremendously, so that now
ships are of welded construction. This suggests that the
advantages in the use of welding exceed the disadvantages.

Welded construction is much lighter than the equivalent riveted construction, due mainly to the reduction in overlaps and flanges. This means that a welded ship may carry more cargo on the same load draught. Welding, if properly carried out, is always watertight without necessitating caulking, while in service riveted joints may readily leak. With the reduction in overlaps, the structure of the ship is much smoother. This leads to reduction in hull resistance and hence the fuel consumption, particularly in the first few years of the ship's life. The smoother surface is easier to clean and less susceptible to corrosion. This is of primary importance in the case of oil tankers where the change from riveting to welding was very rapid. A welded joint is stronger than the equivalent riveted joint, leading to a stronger ship.

Unfortunately a faulty weld may prove much more dangerous than poor riveting, and at the same time is more difficult to detect. The methods of testing welded joints given below, are, while quite successful, nevertheless expensive. If a crack starts in a plate it will, under stress, pass through the plate until it reaches the edge. In riveted construction the edges are common and hence the crack does not have serious results. In welded construction, however, the plates are continuous and hence such a crack may prove very dangerous. It is therefore necessary in welded ships to provide a number of longitudinal *crack arrestors* in the main hull structure to reduce the effects of transverse cracks. These crack arrestors may be in the form of riveted seams or strakes of *extra notch tough* steel through which a crack will not pass. At the same time, great care must be taken in the design of the structure to reduce the possibility of such cracks, by rounding the corners of openings in the structure and by avoiding concentrations of weld metal. It must be clearly understood, however, that if the cracks appear due to inherent weakness of the ship, *i.e.*, if the bending moment creates unduly high stresses, the crack will pass through the plates whether the ship is riveted, welded or a combination of each.

Testing of welds.

There are two basic types of test carried out on welded joints (a) destructive tests and (b) non-destructive tests.

(a) *Destructive tests*

As the heading implies, specimens of the weld material or welded joint are tested until failure occurs, to determine their maximum strength. The tests are those which are used for any metals:

   (i) a tensile test in which the mean tensile strength must be at least 400 MN/m².

   (ii) a bend test in which the specimen must be bent through an angle of 90° with an internal radius of 4 times the thickness of the specimen, without cracking at the edges.

   (iii) an impact test in which the specimen must absorb at least 47 J at about 20°C.

   (iv) any deep penetration electrodes must show the extent of penetration by cutting through a welded section and etching the outline of the weld by means of dilute hydrochloric acid. This test may be carried out on any form of welded joint.

Types of electrode, plates and joints may be tested at regular intervals to ensure that they are maintained at the required standard, while new materials may be checked before being issued for general use. The destructive testing of production work is very limited since it simply determines the strength of the joint before it was destroyed by the removal of the test piece;

Non-destructive tests
Visual inspection of welded joints is most important in order to ensure that there are no obvious surface faults such as cracks and undercut, and to check the leg length and throat thickness of fillet welds.

For internal inspection of shipyard welds, *radiography* is used in the form of X-rays or gamma rays, the former being the most common. Radiographs are taken of important butt welds by passing the rays through the plate onto a photographic plate. Any differences in the density of the plate allow greater exposure of the plate and may be readily seen when developed. Such differences are caused by faults which have the effect of reducing the thickness of the plate. In way of such faults it is necessary to take X-rays at two angles. The resultant films are inserted in a stereoscope which gives the illusion of the third dimension. It is not possible to test fillet welds by means of radiography. It is usual to take 400 to 500 X-rays of welded joints, checking highly stressed members, joints in which cracks are common, and work carried out by different welders on the ship.

Other non-destructive tests are available but are not common in shipbuilding. Surface cracks which are too fine to see even with the aid of a magnifying glass, may be outlined with the aid of a fluorescent penetrant which enters the crack and may be readily seen with the aid of ultra-violet light.

Faults at or near the surface of a weld may be revealed by means of *magnetic crack detection*. An oil containing particles of iron is poured over the weld. A light electric current is passed through the weld. In way of any surface faults a magnetic field will be set up which will create an accumulation of the iron particles. Since the remainder of the iron remains in the oil which runs off, it is easy to see where such faults occur.

A more modern system which is being steadily established is the use of *ultrasonics*. A high frequency electric current causes a quartz crystal to vibrate at a high pitch. The vibrations are transmitted directly through the material being tested. If the material is homogeneous, the vibration is reflected from the opposite surface, converted to an electrical impulse and indicated on an oscilloscope. Any fault in the material, no matter how small, will cause an intermediate reflection which may be noted on the screen. This method is useful in that it will indicate a lamination in a plate which will not be shown on an X-ray plate. Ultrasonics are now being used to determine the thickness of plating in repair work and avoiding the necessity of drilling through the plate.

Faults in welded joints

Electric welding, using correct technique, suitable materials and conditions, should produce faultless welds. Should these requirements not be met, however, faults will occur in the joint. If the current is too high the edge of the plate may be burned away. This is known as *undercut* and has the effect of reducing the thickness of the plate at that point. It is important to chip off all of the slag, particularly in multi-run welds, otherwise *slag inclusions* occur in the joint, again reducing the effective thickness of the weld. The type of rod and the edge preparation must be suitable to ensure *complete penetration* of the joint. In many cases a good surface appearance hides the lack of fusion beneath, and, since this fault may be continuous in the weld, could prove very dangerous. Incorrect welding technique sometimes causes bubbles of air to be trapped in the weld. These bubbles tend to force their way to the surface leaving *pipes* in the weld. Smaller bubbles in greater quantities are known as *porosity*. *Cracks* on or below the surface may occur due to unequal cooling rates or an accumulation of weld metal. The rate of cooling is also the cause of distortion in the plates, much of which may be reduced by correct welding procedure.

Another fault which is attributed to welding but which may occur in any thick plate, especially at extremely low temperatures, is *brittle fracture*. Several serious failures occurred during and just after World War II, when large quantities of welded work were produced. Cracks may start at relatively small faults and suddenly pass through the plating at comparatively small stresses. It is important to ensure that no faults or discontinuities occur, particularly in way of important structural members. The grade of steel used must be suitable for welding, with careful control of the manganese/carbon content in the greater thicknesses to ensure notch-tough qualities.

Design of welded structure

It is essential to realise that welding is different from riveting not only as a process, but as a method of attachment. It is not sufficient to amend a riveted structure by welding, the structure, and indeed the whole shipyard, must be designed for welding. Greater continuity of material may be obtained than with riveting, resulting in more efficient designs. Many of the faults which occurred in welded ships were due to the large number of members which were welded together with resulting high stress points. Consider the structure of an oil tanker. Fig. 3.24 shows part of a typical riveted centre girder, connected to a vertical bulkhead web.

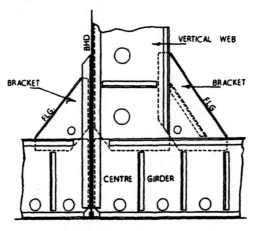

Fig. 3.24

When such ships were built of welded construction, the same type of design was used, the riveting being replaced by welding, resulting in the type of structure shown in Fig. 3.25.

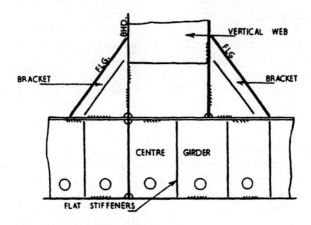

Fig. 3.25

It was found with such designs that cracks occurred at the toes of the brackets and at the ends of the flats. The brackets were then built in to the webs using continuous face flats having small radii at the toes of the brackets. Cracks again appeared showing that the curvature was too small. The radii were increased until

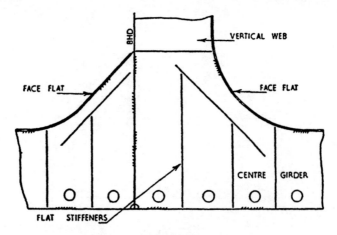

Fig. 3.26

eventually the whole bracket was formed by a large radius joining the bottom girder to the vertical web (Fig. 3.26). This type of structure is now regarded as commonplace in oil tanker design.

Great care must be taken to ensure that structural members on opposite sides of bulkheads are perfectly in line, otherwise cracks may occur in the plating due to shearing.

In the succeeding chapters dealing with ship construction, welded structure is shown where it is most prevalent.

MATERIALS

Mild steel

Mild steel or low carbon steel in several grades has been used as a ship structural material for over a century. It has the advantage of having a relatively good strength-weight ratio, whilst the cost is not excessive.

There are four grades of steel in common use, specified by the Classification Societies as Grades A, B, D and E depending largely upon their degree of notch toughness. Grade A has the least resistance to brittle fracture whilst Grade E is termed 'extra notch tough'. Grade D has sufficient resistance to cracks for it to be used extensively for main structural material.

The disposition of the grades in any ship depends upon the thickness of the material, the part of the ship under consideration and the stress to which it may be subject. For instance, the bottom shell plating of a ship within the midship portion of the ship will have the following grade requirements.

Plate thickness	Grade of steel
Up to 20.5 mm	A
20.5 to 25.5 mm	B
25.5 mm to 40 mm	D
Above 40 mm	E

The tensile strength of the different grades remains constant at between 400 MN/m² and 490MN/m². The difference lies in the chemical composition which improves the impact strength of D and E steels. Impact resistance is measured by means of a Charpy test in which specimens may be tested at a variety of temperatures. The following table shows the minimum values required by Lloyd's Register.

Type of steel	Temperature	Impact resistance
B	0°C	27 joules
D	0°C	47 joules
E	−40°C	27 joules

Higher tensile steels

As oil tankers and bulk carriers increased in size the thickness of steel required for the main longitudinal strength members also increased. In an attempt to reduce the thickness of material and hence reduce the light displacement of the ship, Classification Societies accept the use of steels of higher tensile strength. These steels are designated AH, BH, DH and EH and may be used to replace the normal grades for any given structural member. Thus a bottom shell plate amidships may be 30 mm in thickness of grade DH steel.

The tensile strength is increased to between 490 MN/m^2 and 620 MN/m^2, having the same percentage elongation as the low carbon steel. Thus it is possible to form a structure combining low carbon steel with the more expensive, but thinner higher tensile steel. The latter is used where it is most effective, *i.e.*, for upper deck plating and longitudinals, and bottom shell plating and longitudinals.

Care must be taken in the design to ensure that the hull has an acceptable standard of stiffness, otherwise the deflection of the ship may become excessive. Welding must be carried out using low hydrogen electrodes, together with a degree of preheating. Subsequent repairs must be carried out using the same type of steel and electrodes. It is a considerable advantage if the ship carries spare electrodes, whilst a plan of the ship should be available showing the extent of the material together with its specification.

Arctic D steel

If part of the structure of a ship is liable to be subject to particularly low temperatures, then the normal grades of steel are not suitable. A special type of steel, known as Arctic D, has been developed for this purpose. It has a higher tensile strength than normal mild steel, but its most important quality is its ability to absorb a minimum of 40 J at −55°C in a Charpy impact test using a standard specimen.

Aluminium alloys

Pure aluminium is too soft for use as a structural material and must be alloyed to provide sufficient strength in relation to the mass of material used. The aluminium is combined with copper, magnesium, silicon, iron, manganese, zinc, chromium and titanium, the manganese content varying between about 1% and 5% depending upon the alloy. The alloy must have a tensile strength of 260 MN/m^2 compared with 400 MN/m^2 to 490 MN/m^2 for mild steel.

There are two major types of alloy used in shipbuilding, heat-treatable and non-heat-treatable. The former is heat-treated during manufacture and, if it is subsequently heat-treated, tends to lose its strength. Non-heat-treatable alloys may be readily welded and subject to controlled heat treatment whilst being worked.

The advantages of aluminium alloy in ship construction lie in the reduction in weight of the material and its non-magnetic properties. The former is only important, however, if sufficient material is used to significantly reduce the light displacement of the ship and hence increase the available deadweight or reduce the power required for any given deadweight and speed. Unfortunately, the melting point of the alloy (about 600°C) lies well below the requirements of a standard fire test maximum temperature (927°C). Thus if it is to be used for fire subdivisions it must be suitably insulated.

The major application of aluminium alloys as a shipbuilding material is in the construction of passenger ships, where the superstructure may be built wholly of the alloy. The saving in weight at the top of the ship reduces the necessity to carry permanent ballast to maintain adequate stability. The double saving results in an economical justification for the use of the material. Great care must be taken when attaching the aluminium superstructure to the steel deck of the main hull structure (see Chap. 12).

Other applications in passenger ships have been for cabin furniture, lifeboats and funnels.

One tremendous advantage of aluminium alloy is its ability to accept impact loads at extremely low temperatures. Thus it is an eminently suitable material for main tank structure in low-temperature gas carriers.

Brittle fracture

When welding was first introduced into shipbuilding on an extensive scale, several structural failures occurred. Cracks were found in ships which were not highly stressed, indeed in some cases the estimated stress was particularly low. On investigation it was found that the cracks were of a brittle nature, indicated by the crystalline appearance of the failed material. Further study indicated that similar types of fault had occurred in riveted ships although their consequences were not nearly as serious as with welded vessels.

Series of tests indicated that the failures were caused by brittle fracture of the material. In some cases it was apparent that the crack was initiated from a notch in the plate; a square corner on an opening or a fault in the welding. (In 1888 Lloyd's Register pointed out the dangers of square corners on openings.) At other times cracks appeared suddenly at low temperatures whilst the stresses were particularly low and no structural notches appeared in the area. Some cracks occurred in the vicinity of a weld and were attributed to the change in the composition of the steel due to welding. Excessive impact loading also created cracks with a crystalline appearance. Explosions near the material caused dishing of thin plate but cracking of thick plate.

The consequences of brittle fracture may be reduced by fitting crack arrestors to the ship where high stresses are likely to occur. Riveted seams or strakes of extra notch tough steel are fitted in the decks and shell of large tankers and bulk carriers.

Brittle fracture may be reduced or avoided by designing the structure so that notches in plating do not occur, and by using steel which has a reasonable degree of notch-toughness. Grades D and E steel lie in this category and have proved very successful in service for the main structure of ships where the plates are more than about 12 mm thick.

CHAPTER 4

BOTTOM AND SIDE FRAMING

DOUBLE BOTTOM

All ocean-going ships with the exception of tankers, and most coastal vessels are fitted with a double bottom which extends from the fore peak bulkhead almost to the after peak bulkhead.

The double bottom consists of the outer shell and an inner skin or tank top between 1 m and 1.5 m above the keel. This provides a form of protection in the event of damage to the bottom shell. The tank top, being continuous, increases the longitudinal strength and acts as a platform for cargo and machinery. The double bottom space contains a considerable amount of structure and is therefore useless for cargo. It may, however, be used for the carriage of oil fuel, fresh water and water ballast. It is sub-divided longitudinally and transversely

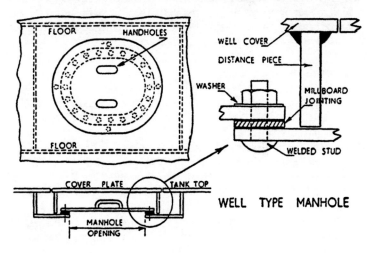

Fig. 4.1

into large tanks which allow different liquids to be carried and may be used to correct the heel of a ship or to change the trim. Access to these tanks is arranged in the form of manholes with watertight covers (Fig. 4.1).

In the majority of ships only one watertight longitudinal division, a centre girder, is fitted, but many modern ships are designed with either three or four tanks across the ship. A cofferdam must be fitted between a fuel tank and a fresh water tank to prevent contamination of one with the other. The tanks are tested by pressing them up until they overflow. Since the overflow pipe usually extends above the weather deck, the tank top is subject to a tremendous head which in most cases will exceed the load from the cargo in the hold. The tank top plating must be thick enough to prevent undue distortion. If it is anticipated that cargo will be regularly discharged by grabs or by fork lift trucks, it is necessary to fit either a double wood ceiling or heavier flush plating. Under hatchways, where the tank top is most liable to damage, the plating must be increased or wood ceiling fitted. The plating is 10% thicker in the engine room.

At the bilges the tank top may be either continued straight out to the shell, or knuckled down to the shell by means of a *tank margin* plate set at an angle of about 45° to the tank top and meeting the shell almost at right angles. This latter system was originally used in riveted ships in order to obtain an efficient, watertight connection between the tank top and the shell. It has the added advantage, however, of forming a bilge space into which water may drain and proves to be most popular. If no margin plate is fitted it is necessary to fit drain hats or wells in the after end of the tank top in each compartment.

Internal structure

A continuous *centre girder* is fitted in all ships, extending from the fore peak to after peak bulkhead. This girder is usually watertight except at the extreme fore and after ends where the ship is narrow, although there are some designs of ship where the centre girder does not form a tank boundary and is therefore not watertight. Additional longitudinal *side girders* are fitted depending upon the breadth of the ship but these are neither continuous nor watertight, having large manholes or lightening holes in them.

The tanks are divided transversely by *watertight floors* which in most ocean-going ships, are required to be stiffened vertically to withstand the liquid pressure. Fig. 4.2 shows a typical, welded watertight floor.

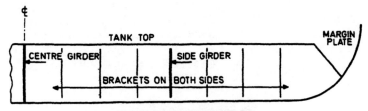

WATERTIGHT FLOOR

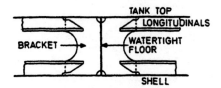

END CONNECTIONS
OF LONGITUDINALS

Fig. 4.2

In ships less than 120 m in length the bottom shell and tank top are supported at intervals of not more than 3 m by transverse plates known as *solid floors* (Fig. 4.3). The name slightly belies the structure since large lightening holes are cut in them. In addition, small air release and drain holes are cut at the top and bottom respectively. These holes are most important since it is essential to have adequate access and ventilation to all parts of the double bottom. There have been many cases of personnel entering tanks which have been inadequately ventilated, with resultant gassing or suffocation.

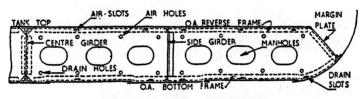

SOLID FLOOR — RIVETED

Fig. 4.3

The solid floor is usually fitted as a continuous plate extending from the centre girder to the margin plate. The side girder is therefore broken on each side of the floor plate and is said to be *intercostal*.

Solid floors are required at every frame space in the machinery room, in the forward quarter length and elsewhere where heavy loads are experienced, such as under bulkheads and boiler bearings.

The remaining bottom support may be of two forms:

(a) transverse framing.

(b) longitudinal framing.

Transverse framing has been used for the majority of riveted ships and for many welded ships. The shell and tank top between the widely-spaced solid floors are stiffened by bulb angles or similar sections running across the ship and attached at the centreline and the margin to large flanged brackets. Additional support is given to these stiffeners by the side girder and by intermediate struts which are fitted to reduce the span. Such a structure is known as a bracket floor (Fig. 4.4).

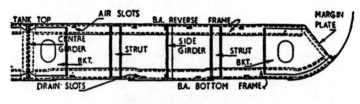

BRACKET FLOOR

Fig. 4.4

It was found that the distortion due to the welding of the floors and frames, together with the bending of the ship, caused the corrugation of the bottom shell, which, in many welded ships, assumed dangerous proportions. While there are no records of ship losses due to this fault, many ships were required to fit short longitudinal stiffeners. Such deflections were reduced in riveted ships by the additional stiffening afforded by the flanges of the angles and by the longitudinal riveted seams. This problem was overcome by using longitudinal stiffening in the double bottom of welded ships, a system recommended for all ships over 120 m long. Longitudinal frames are fitted to the bottom shell and under the tank top, at intervals of about 760 mm. They are supported by the solid floors mentioned earlier, although the spacing of these floors may be increased to 3.7 m. Intermediate struts are fitted so that the unsupported span of the longitudinals does not exceed 2.5 m. Brackets are again required at the margin plate and centre girder, the latter being necessary when docking. Fig. 4.5 shows an arrangement of double bottom in a welded ship.

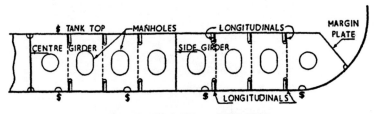

SOLID FLOOR — WELDED

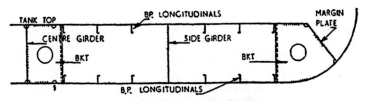

LONGITUDINAL FRAMING

Fig. 4.5

The longitudinals are arranged to line up with any additional longitudinal girders which are required for machinery support in the engine room.

Duct keel

Some ships are fitted with a duct keel which extends from within the engine room length to the forward hold. This arrangement allows pipes to be carried beneath the hold spaces and are thus protected against cargo damage. Access into the duct is arranged from the engine room, allowing the pipes to be inspected and repaired at any time. At the same time it is possible to carry oil and water pipes in the duct, preventing contamination which could occur if the pipes passed through tanks. Duct keels are particularly important in insulated ships, allowing access to the pipes without disturbing the insulation. Ducts are not required aft since the pipes may be carried through the shaft tunnel.

The duct keel is formed by two longitudinal girders up to 1.83 m apart. This distance must not be exceeded as the girders must be supported by the keel blocks when docking. The structure on each side of the girders is the normal double bottom arrangement. The keel and the tank top centre strake must be strengthened either by supporting members in the duct or by increasing the thickness of the plates considerably.

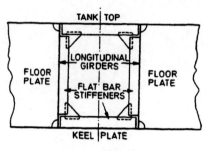

DUCT KEEL

Fig. 4.6

Double bottom in the machinery space

Great care must be taken in the machinery space to ensure that the main and auxiliary machinery are efficiently supported. Weak supports may cause damage to the machinery, while large unsupported panels of plating may lead to vibration of the structure. The main engine bedplate is bolted through a tank top plate which is about 40 mm thick and is continuous to the thrust block seating. A girder is fitted on each side of the bedplate in such a way that the holding down bolts pass through the top angle of the girder. In welded ships a horizontal flat is sometimes fitted to the top of the girder in way of the holding-down bolts (Fig. 4.7).

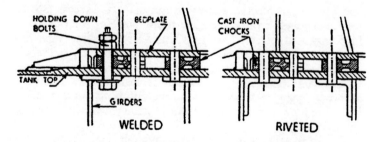

MAIN — ENGINE HOLDING DOWN BOLTS

Fig. 4.7

In motor ships where a drain tank is required under the machinery, a cofferdam is fitted giving access to the holding down bolts and isolating the drain from the remainder of the double bottom tanks. Additional longitudinal girders are fitted in way of heavy auxiliary machinery such as generators.

SIDE FRAMING

The side shell is supported by frames which run vertically from the tank margin to the upper deck. These frames, which are spaced about 760 mm apart, are in the form of bulb angles and channels in riveted ships or bulb plates in welded ships. The lengths of frames are usually broken at the decks, allowing smaller sections to be used in the 'tween deck spaces where the load and span are reduced. The hold frames are of large section (300 mm bulb angle). They are connected at the tank margin to flanged tank side brackets (Fig. 4.8). To prevent the free edge of the brackets buckling, a gusset plate is fitted, connecting the flange of the brackets to the tank top. A hole is cut in each bracket to allow the passage of bilge pipes. In insulated ships the tank top may be extended to form the gusset plate and the tank side bracket fitted below the level of the tank top (Fig. 4.9). This increases the cargo capacity and facilitates the fitting of the insulation. Since the portion of the bracket above the tank top

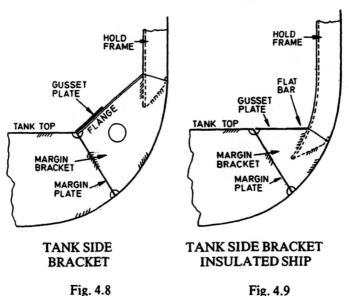

TANK SIDE BRACKET	TANK SIDE BRACKET INSULATED SHIP
Fig. 4.8	Fig. 4.9

level is dispensed with, the effective span of the frame is increased, causing an increase in the size of the frame.

The top of the hold frames terminate below the lowest deck and are connected to the deck by beam knees (Fig. 4.10) which may be flanged on their free edge. The bottom of the 'tween deck frames are usually welded directly to the deck, the deck plating at the side being knuckled up to improve drainage. At the top, the 'tween deck frames are stopped slightly short of the upper deck and connected by beam knees (Fig. 4.11).In some cases the 'tween deck frames must be carried through the second deck and it is necessary to fit a collar round each frame to ensure that the deck is watertight. Fig. 4.12 shows a typical collar arrangement, the collar being in two pieces, welded right round the edges.

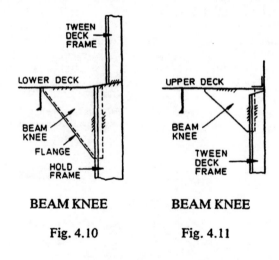

BEAM KNEE

Fig. 4.10

BEAM KNEE

Fig. 4.11

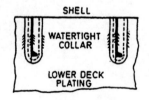

WELDED PLATE
COLLARS

Fig. 4.12

Wood sparring is fitted to the toes of the hold and 'tween deck frames to protect the cargo from damage, while the top of the tank side brackets in the holds are fitted with wood ceiling.

Web frames are fitted in the machinery and connected to strong beams or pillars in an attempt to reduce vibration. These web frames are about 600 mm deep and are stiffened on their free edge. It is usual to fit two or three web frames on each side of the ship, a smaller web being fitted in the 'tween decks.

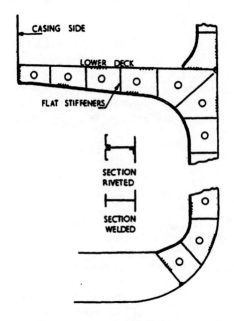

WEB FRAME

Fig. 4.13

CHAPTER 5

SHELL AND DECKS

The external hull of a ship consists of bottom shell, side shell and decks which are formed by longitudinal strips of plating known as *strakes*. The strakes themselves are constructed of a number of plates joined end to end. Large, wide plates should be used to reduce the welding required but are usually restricted by transport difficulties and limitations of shipyard machinery.

SHELL PLATING

The bottom and side shell plating of a ship form a major part of the longitudinal strength members of the vessel. The most important part of the shell plating is that on the bottom of the ship, since this is the greatest distance from the neutral axis. It is therefore slightly thicker than the side shell plating. The *keel plate* is about 30% thicker than the remainder of the bottom shell plating, since it is subject to wear when docking. The strake adjacent to the keel on each side of the ship is known as the *garboard strake* which is the same thickness as the remainder of the bottom shell plating. The uppermost line of plating in the side shell is known as the *sheerstrake* which is 10% to 20% thicker than the remaining side shell plating.

The thickness of the shell plating depends mainly on the length of the ship, varying between about 10 mm at 60 m to 20 mm at 150 m. The depth of the ship, the maximum draught and the frame spacing are, however, also taken into account. If the depth is increased it is possible to reduce the thickness of the plating. In ships fitted with long bridges which extend to the sides of the ship, the depth in way of the bridge is increased, resulting in thinner shell plating. Great care must be taken at the ends of such superstructures to ensure that the bridge side plating is tapered gradually to the level of the upper deck, while the thicker shell plating forward and aft of the bridge must be taken past the ends of the bridge to form an efficient scarph. If the draught of the ship is increased, then the shell plating must also be increased. Thus a ship whose freeboard is measured from the upper deck has thicker shell plating than a similar ship whose freeboard is measured from the second deck. If the frame spacing is increased the shell plating is required to be increased. The maximum bending moment of a ship occurs at or near amidships. Thus it is reasonable to build the ship stronger amidships than at the ends. The main shell plating has its thickness maintained 40% of its length amidships and tapered *gradually* to a minimum thickness at the ends of the ship.

While the longitudinal strength of shell plating is of prime importance, it is equally important that its other functions are not overlooked. Watertight hulls were made before longitudinal strength was considered. It is essential that the shell plating should be watertight, and, at the same time, capable of withstanding the static and dynamic loads created by the water. The shell plating, together with the frames and double bottom floors, resist the water pressure, while the plating must be thick enough to prevent undue distortion between the frames and floors. If it is anticipated that the vessel will regularly travel through ice, the shell plating in the region of the waterline forward is increased in thickness and small intermediate frames are fitted to reduce the widths of the panels of plating. The bottom shell plating forward is increased in thickness to reduce the effects of pounding (see Chapter 7).

The shell plating and side frames act as pillars supporting the loads from the decks above and must be able to withstand the weight of the cargo. In most cases the strength of the panel which is required to withstand the water pressure is more than sufficient to support the cargo, but where the internal loading is particularly high, such as in way of a deep tank, the frames must be increased in strength.

It is necessary on exposed decks to fit some arrangement to prevent personnel falling or being washed overboard. Many ships are fitted with open rails for this purpose while others are fitted with solid plates known as *bulwarks* at least 1 m high. These bulwarks are much thinner than the normal shell plating and are not regarded as longitudinal strength members. The upper edge is stiffened by a 'hooked angle,' *i.e.,* the plate is fitted inside the flange. This covers the free edge of the plate and results in a neater arrangement. Substantial stays must be fitted from the bulwark to the deck at intervals of 1.83 m or less. The lower edge of the bulwark in riveted ships is riveted to the top edge of the sheerstrake. In welded ships, however, there must be no direct connection between the bulwark and the sheerstrake, especially amidships, since the high stresses would then be transmitted to the bulwark causing cracks to appear. These cracks could then pass through the sheerstrake. Large openings, known as *freeing ports,* must be cut in the bottom of the bulwark to allow the water to flow off deck when a heavy sea is shipped. Failure to clear the water could cause the ship to capsize. Rails or grids are fitted to restrict the opening to 230 mm in depth, while many ships are fitted with hinged doors on the outboard side of the freeing port, acting as rather inefficient non-return valves. It is essential that there should be no means of bolting the door in the closed position.

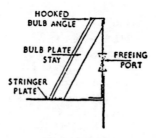

BULWARK

Fig. 5.1

DECK PLATING

The deck plating of a ship carries a large proportion of the stressses due to longitudinal bending, the upper deck carrying greater loads than the second deck. The continuous plating alongside the hatches must be thick enough to withstand the loads. The plating between the hatches has little effect on the longitudinal strength. The thickness of plating depends largely upon the length of the ship and the width of deck alongside the hatchways. In narrow ships, or in vessels having wide hatches, the thickness of plating is increased. At the ends of the ship, where the bending moments are reduced, the thickness of plating may be gradually reduced in the same way as the shell plating. A minimum cross sectional area of material alongside hatches must be maintained. Thus if part of the deck is cut away for a stairway or similar opening, compensation must be made in the form of either doubling plates or increased local plate thickness.

The deck forms a cover over the cargo, accommodation and machinery space and must therefore be watertight. The weather deck, and usually the second deck, are cambered to enable water to run down to the sides of the ship and hence overboard through the scuppers. The outboard deck strake is known as the *stringer plate* and at the weather deck is usually thicker than the remaining deck plating. It may be connected to the sheerstrake by means of a continuous *stringer angle* or *gunwhale bar*.

Exposed steel decks above accommodation must be sheathed with wood which acts as heat and sound insulation. As an alternative the deck may be covered with a suitable composition. The deck must be adequately protected against corrosion between the steel and the wood or composition. The deck covering is stopped short of the sides of the deck to form a waterway to aid drainage.

BEAMS AND DECK GIRDERS

The decks may be supported either by transverse beams in conjunction with longitudinal girders or by longitudinal beams in conjunction with transverse girders.

The transverse beams are carried across the ship and bracketed to the side frames by means of *beam knees*. A continuous longitudinal girder is fitted on each side of the ship alongside the hatches. The beams are bracketed or lugged to the girders, thus reducing their span. In way of the hatches, the beams are broken to allow open hatch space, and are joined at their inboard ends to either the girder or the hatch side coaming. A similar arrangement is necessary in way of the machinery casings. These broken beams are known as *half beams*. The beams are usually bulb angles in riveted ships and bulb plates in welded ships.

There are several forms of girder in use, some of which are shown in Fig. 5.2.

If the girder is required to form part of the hatch coaming, the flanged girder Fig. 5.2 (*i*) is most useful since it is easy to produce and does not require the addition of a moulding to prevent chafing of ropes. Symmetrical girders such as Fig. 5.2 (*iii*) are more efficient but cannot form part of a hatch side coaming. Such girders must be fitted outboard of the hatch sides. The girders are bracketed to the transverse bulkheads and are supported at the hatch corners either by pillars or by hatch

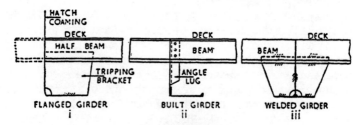

DECK GIRDERS

Fig. 5.2

end girders extending right across the ship. Tubular pillars are most often used in cargo spaces since they give utmost economy of material and, at the same time, reduce cargo damage. In deep tanks, where hollow pillars should not be used, and in machinery spaces, either built pillars or broad flanged beams prove popular.

Most modern ships are fitted with longitudinal beams which extend, as far as practicable, along the whole length of the ship outside the line of the hatches. They are bracketed to the transverse bulkheads and are supported by transverse girders which are carried right across the ship, or, in way of the hatches and machinery casings, from the side of the ship to the hatch or casing. The increase in continuous longitudinal material leads to a reduction in deck thickness. The portion of deck between the hatches may be supported either by longitudinal or transverse beams, neither having any effect on the longitudinal strength of the ship.

At points where concentrated loads are anticipated it is necessary to fit additional deck stiffening. Additional support is required in way of winches, windlasses and capstans. The deck machinery is bolted to seatings which may be riveted or welded to the deck. The seatings are extended to distribute the load. In way of the seatings, the beams are increased in strength by fitting reverse bars which extend to the adjacent girders. Solid pillars are fitted under the seatings to reduce vibration.

WINCH SEATING

Fig. 5.3

HATCHES

Large hatches must be fitted in the decks of dry cargo ships to facilitate loading and discharging of cargo. It is usual to provide one hatch per hold or 'tween deck, although in ships having large holds two hatches are sometimes arranged. The length and width of hatch depend largely upon the size of the ship and the type of cargo likely to be carried. General cargo ships have hatches which will allow cargoes such as timber, cars, locomotives and crates of machinery to be loaded. A cargo tramp of about 10 000 tonne deadweight may have five hatches, each 10 m long and 7 m wide, although one hatch, usually to No. 2 hold, is often increased in length. Large hatches also allow easy handling of cargoes. Bulk carriers have long, wide hatches to allow the cargo to fill the extremities of the compartment without requiring trimming manually.

The hatches are framed by means of hatch coamings which are vertical webs forming deep stiffeners. The heights of the coamings are governed by the Load Line Rules. On weather decks they must be at least 600 mm in height at the fore end and either 450 mm or 600 mm aft depending upon the draught of the ship. Inside superstructures and on lower decks no particular height of coaming is specified. It is necessary, however, for safety considerations, to fit some form of rail around any deck opening to a height of 800 mm. It is usual, therefore, at the weather deck, to extend the coaming to a height of 800 mm. In the superstructures and on lower decks portable stanchions are provided, the rail being in the form of a wire rope. These rails are only erected when the hatch is opened.

The weather deck hatch coamings must be 11 mm thick and must be stiffened by a moulding at the top edge. Where the height of the coaming is 600 mm or more, a horizontal bulb angle or bulb plate is fitted to stiffen the coaming which has additional support in the form of stays fitted at intervals of 3 m. Fig. 5.4 gives a typical section through the side coaming of a weather deck hatch. The edge stiffening is in the form of a bulb angle set back from the line of the coaming. This forms a rest to support the portable beams. The edge stiffening on the hatch end coaming (Fig. 5.5) is a Tyzack moulding which is designed to carry the ends of the wood boards.

The hatch coamings inside the superstructures are formed by 230 mm bulb angles or bulb plates at the sides and ends. The side coamings are usually set back from the opening to form a beam rest, while an angle is fitted at the ends to form a rest bar for the ends of the wood covers.

The hatches may be closed by wood boards which are supported by the portable hatch beams. The beams may be fitted in guides attached to the coamings and lifted out to clear the hatch, or fitted with rollers allowing them to be pushed to the hatch ends. The covers are made weathertight by means of tarpaulins which are wedged tight at the sides and ends (Fig. 5.6), at least two tarpaulins being fitted on weather deck hatches.

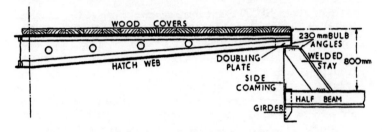

TRANSVERSE SECTION THROUGH HATCH
Fig. 5.4

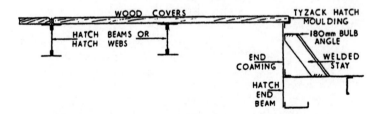

LONGITUDINAL SECTION THROUGH HATCH
Fig. 5.5

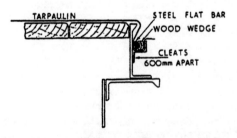

HATCH CLOSING ARRANGEMENT
Fig. 5.6

Modern ships are fitted with steel hatch covers. There are many types available, from small pontoons supported by portable beams to the larger self-supporting type, the latter being the most popular. The covers are arranged in four to six sections extending right across the hatch and having rollers which rest on a runway. The covers are opened by rolling them to the end of the hatch where they tip automatically into the vertical position. The separate sections are joined by means of wire rope, allowing opening or closing to be a continuous action, a winch being used for the purpose.

Many other systems are available, some with electric or hydraulic motors driving sprocket wheels, some in which the whole cover wraps round a powered drum, whilst others have hydraulic cylinders built into the covers. In the latter arrangement pairs of covers are hinged together, the pairs being linked to provide continuity. Each pair of covers has one or two hydraulic rams which turn the hinge through 180°. The rams are actuated by an external power source, with a control panel on the side of the hatch coaming.

The covers interlock at their ends and are fitted with packing to ensure that when the covers are wedged down, a watertight cover is provided (Fig. 5.8). Such covers do not require tarpaulins. At the hatch sides the covers are held down by cleats which may be manual as shown in Fig.5.9 or hydraulically operated.

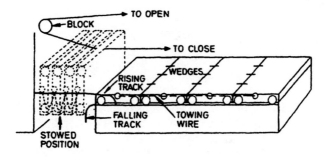

STEEL HATCH COVERS

Fig. 5.7

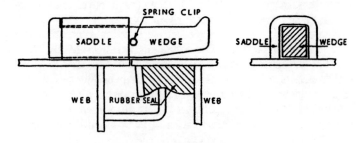

HATCH WEDGES

Fig. 5.8

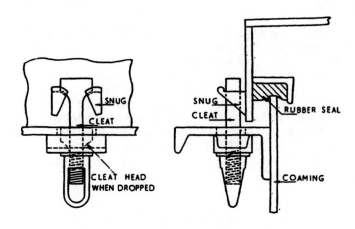

SECURING CLEATS

Fig. 5.9

Deep tank hatches have two functions to fulfil. They must be watertight or oiltight and thus capable of withstanding a head of liquid, and they must be large enough to allow normal cargoes to be loaded and discharged if the deep tank is required to act as a dry cargo hold. Such hatches may be 3 m or 4 m square. Because of the possible liquid pressure, the covers must be stiffened, while some suitable packing must be fitted in the coamings to ensure watertightness, together with some means of securing the cover. The covers may be hinged or may be arranged to slide.

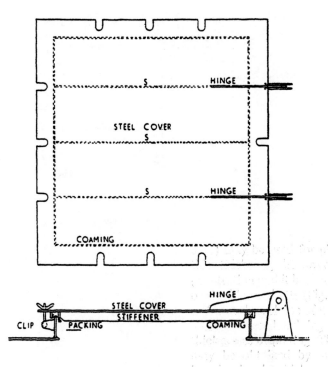

DEEP TANK HATCH

Fig. 5.10

Fig. 5.11 summarises much of the foregoing work by showing the relation between the separate parts in a welded ship. The sizes or scantlings of the structure are suitable for a ship of about 10 000 tonne deadweight.

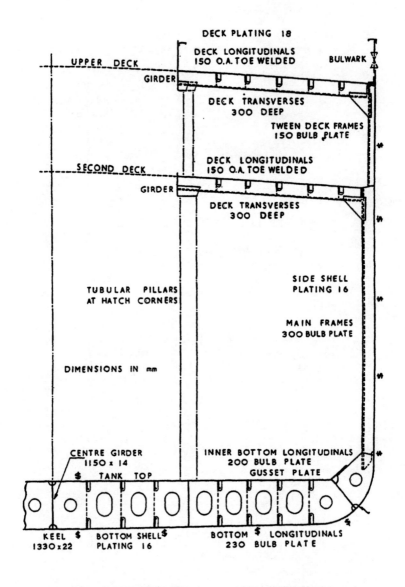

MIDSHIP SECTION   WELDED SHIP

Fig. 5.11

CHAPTER 6

BULKHEADS AND DEEP TANKS

There are three basic types of bulkheads used in ships; watertight bulkheads, tank bulkheads and non-watertight bulkheads. These bulkheads may be fitted longitudinally or transversely, although only non-watertight and some tank bulkheads are fitted longitudinally in most dry cargo ships.

WATERTIGHT BULKHEADS

The transverse watertight bulkheads of a ship have several functions to perform.

They divide the ship into watertight compartments and thus restrict the volume of water which may enter the ship if the shell plating is damaged. In passenger ships, complicated calculations are carried out to ensure an arrangement of bulkheads which will prevent the ship sinking if the ship is damaged to a reasonable extent. A simpler form of calculation is occasionally carried out for cargo ships but results in only a slight indication of the likelihood of the vessel sinking, since the volume and type of cargo play an important role.

The watertight compartments also serve to separate different types of cargo and to divide tanks and machinery spaces from the cargo spaces.

In the event of fire, the bulkheads reduce to a great extent the rate of spread. Much depends upon the fire potential on each side of the bulkhead, *i.e.,* the likelihood of the material near the bulkhead being ignited.

The transverse strength of the ship is increased by the bulkheads which have much the same effect as the ends of a box. They prevent undue distortion of the side shell and reduce racking considerably.

Longitudinal deck girders and deck longitudinals are supported at the bulkheads which therefore act as pillars, while at the same time they tie together the deck and tank top and hence reduce vertical deflection when the compartments are full of cargo.

Thus it appears that the shipbuilder has a very complicated structure to design. In practice, however, it is found that a bulkhead required to withstand a load of water in the event of flooding will readily perform the remaining functions.

The number of bulkheads in a ship depends upon the length of the ship and the position of the machinery space. Each ship must have a collision bulkhead at least one twentieth of the ship's length from the forward perpendicular, which must be continuous up to the uppermost continuous deck. The stern tube must be enclosed in a watertight compartment formed by the sternframe and the after peak bulkhead which may terminate at the first watertight deck above the waterline. A bulkhead must be fitted at each end of the machinery space although, if the engines are aft, the after peak forms the after boundary of the space. In certain ships this may result in the saving of one bulkhead. In ships more than 90 m in length, additional bulkheads are required, the number depending upon the length. Thus a ship 140 m long will require a total of 7 bulkheads if the machinery is amidships or 6 bulkheads if the machinery is aft, while a ship 180 m in length will require 9 or 8 bulkheads respectively. These bulkheads must extend to the freeboard deck and should preferably be equally spaced in the ship. It may be seen, however, from Chapter 1, that the holds are not usually of equal length. The bulkheads are fitted in separate sections between the tank top and the lowest deck, and in the 'tween decks.

Watertight bulkheads are formed by plates which are attached to the shell, deck and tank top by welding (Fig. 6.1). Since water pressure increases with the head, and the bulkhead is to be designed to withstand such a force, it may be expected that the plating on the lower part of the bulkhead is thicker than that at the top. The bulkheads are supported by vertical stiffeners spaced 760 mm apart. Any variation in this spacing results in variations in size of stiffeners and thickness of plating. The ends of the stiffeners are usually bracketed to the tank top and deck although in some cases the brackets are omitted, resulting in heavier stiffeners.

The stiffeners are in the form of either bulb plates or toe welded angles. It is of interest to note that since a welded bulkhead is less liable to leak under load, or alternatively it may deflect further without leakage, the strength of the stiffeners may be reduced by 15%. It may be necessary to increase the strength of a stiffener which is attached to a longitudinal deck girder in order to carry the pillar load.

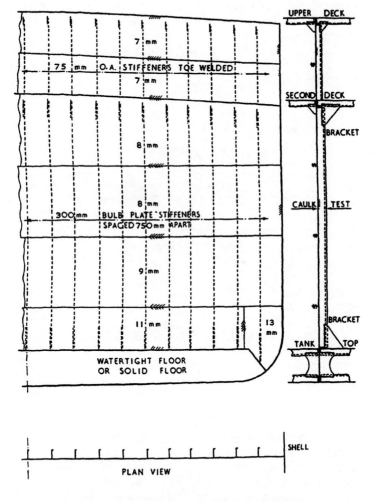

WELDED WATERTIGHT BULKHEAD

Fig. 6.1

The bulkheads are tested for watertightness by hosing them using a pressure of 200 kN/m². The test is carried out from the side on which the stiffeners are attached. It is essential that the structure should be maintained in a watertight condition. If it is found necessary to penetrate the bulkhead, precautions must be taken to ensure that the bulkhead remains watertight. The after engine room bulkhead is penetrated by the main shaft, which passes through a watertight gland, and by an opening leading to the shaft tunnel. This opening must be fitted with a sliding watertight door. When pipes or electric cables pass through a bulkhead, the integrity of the bulkhead must be maintained. Fig. 6.2 shows a bulkhead fitting in the form of a watertight gland for an electric cable.

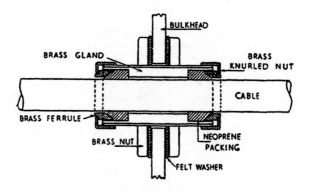

WATERTIGHT CABLE GLAND

Fig. 6.2

In many insulated ships, ducts are fitted to provide efficient circulation of cooled air to the cargo spaces. The majority of such ships are designed so that the ducts from the hold spaces pass vertically through the deck into a fan room, separate rooms being constructed for each hold. In these ships it is not necessary to penetrate any transverse bulkhead with a duct. In some cases, however, it is necessary to penetrate the bulkhead in which case a sliding watertight shutter must be fitted.

WATERTIGHT DOORS

A watertight door is fitted to any access opening in a watertight bulkhead. Such openings must be cut only where necessary for the safe working of the ship and are kept as small as possible, 1.4 m high and 0.75 m wide being usual. The doors may be mild steel, cast steel or cast iron, and either vertical or horizontal sliding, the choice being usually related to the position of any fittings on the bulkhead. The means of closing the doors must be positive, *i.e.,* they must not rely on gravity or a dropping weight.

Vertical sliding doors (Fig. 6.3) are closed by means of a vertical screw thread which turns in a gunmetal nut secured to the door. The screw is turned by a spindle which extends above the bulkhead deck, fitted with a crank handle allowing complete circular motion. A similar crank must be fitted at the door. The door runs in vertical grooves which are tapered towards the bottom, the door having similar taper, so that a tight bearing fit is obtained when the door is closed. Brass facing strips are fitted to both the door and the frame. There must be no groove at the bottom of the door to collect dirt which would prevent the door fully closing. An indicator must be fitted at the control position above the bulkhead deck, showing whether the door is open or closed.

A horizontal sliding door is shown in Fig. 6.4. It is operated by means of an electric motor A which turns a vertical shaft B. Near the top and bottom of the door, horizontal screw shafts C are turned by the vertical shaft through the bevel gears D. The door nut E moves along the screw shaft within the nut box F until any slack is taken up or the spring G is fully compressed, after which the door moves along its wedge-shaped guides on rollers H.

The door may be opened or closed manually at the bulkhead position by means of a handwheel J, the motor being automatically disengaged during this operation. An alarm bell gives warning 10 seconds before the door is to close and whilst it is being closed. Opening and closing limit switches K are built into the system to prevent overloading of the motors.

A de-wedging device (Fig. 6.5) may be fitted to release the door from the wedge frame and to avoid overloading the power unit if the door meets an obstruction. As the door-operating shaft turns, the spring-loaded nut E engages a lever L which comes into contact with a block M on the door frame. As the nut continues to move along the shaft, a force is exerted by the lever

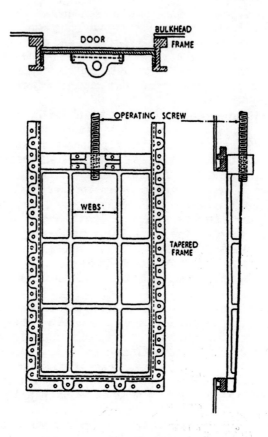

VERTICAL SLIDING
WATERTIGHT DOOR

Fig. 6.3

on the block, easing the door out of the wedge. Should a solid
obstruction be met, the striker N lifts a switch bar P and cuts out
the motor.

A electric motor
B vertical shaft
C horizontal shafts
D bevel gears
E door nut
F nut box
G spring

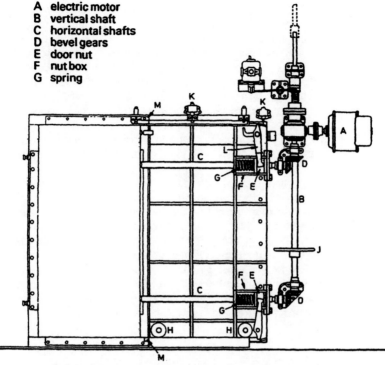

HORIZONTAL SLIDING WATERTIGHT DOORS

Fig. 6.4

H rollers
J handwheel
K limit switches
L de-wedging lever
M de-wedging block
N striker
P switch bar

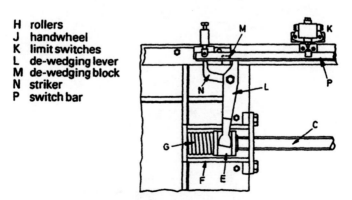

DE-WEDGING DEVICE

Fig. 6.5

Some door systems are hydraulically-operated, having a pumping plant which consists of two units. Each unit is capable of operating all the watertight doors in a passenger ship, the electric motor being connected to an emergency power source. The doors may be closed at the door position or from a control point. If closed from the control point they may be opened from a local position, switches being fitted on both sides of the bulkhead, but close automatically when the switch is released.

Watertight doors for passenger ships are tested before fitting by a hydraulic pressure equivalent to a head of water from the door to the bulkhead deck. All such doors are hose tested after fitting.

Hinged watertight doors may be fitted to watertight bulkheads in passenger ships, above decks which are 2.2 m or more above the load waterline. Similar doors are fitted in cargo ships to weather deck openings which are required to be watertight. The doors are secured by clips which may be fitted to the door or to the frame. The clips are forced against brass wedges. The hinges must be fitted with gunmetal pins. Some suitable packing is fitted round the door to ensure that it is watertight. Fig. 6.6 shows the hinge and clip for a hinged door, six clips being fitted to the frame.

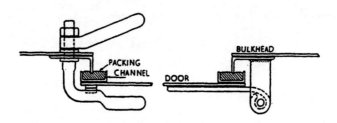

CLIP & HINGE
WATERTIGHT DOOR

Fig. 6.6

DEEP TANKS

It is usually necessary in ships with machinery amidships to arrange a deep tank forward of the machinery space to provide sufficient ballast capacity. This deep tank is usually designed to allow dry cargo to be carried and in many ships may carry vegetable oil or oil fuel as cargo. Deep tanks are also provided for the carriage of oil fuel for use in the ship. The structure in these tanks is designed to withstand a head of water up to the top of the overflow pipe, the tanks being tested to this head or to a height of 2.44 m above the top of the tank, which ever is the greater. It follows, therefore, that the strength of the structure must be much superior to that required for dry cargo holds. If a ship is damaged in way of a hold, the end bulkheads are required to withstand the load of water without serious leakage. Permanent deflection of the bulkhead may be accepted under these conditions and a high stress may be allowed. There must be no permanent deflection of a tank bulkhead, however, and the allowable stress in the stiffeners must therefore be much smaller. The stiffener spacing on the transverse bulkheads is usually about 600 mm and the stiffeners are much heavier than those on hold bulkheads. If, however a horizontal girder is fitted on the bulkhead, the size of the stiffeners may be considerably reduced. The ends of the stiffeners are bracketed, the toe of the bottom bracket being supported by a solid floor plate. The thickness of bulkhead plating is greater than required for hold bulkheads, with a minimum thickness of 7.5mm. The arrangement of the structure depends upon the use to which the tank will be put.

Deep tanks for water ballast or dry cargo only

A water ballast tank should be either completely full or empty while at sea and therefore there should be no movement of water. The side frames are increased in strength by 15% unless horizontal stringers are fitted, when the frames are reduced. If such stringers are fitted, they must be continued across bulkheads to form a ring. These girders are substantial, with stiffened edges. The deck forming the top of a deep tank may be required to be increased in thickness because of the increased load due to water pressure. The beams and deck girders in way of a deep tank are calculated in the same way as the bulkhead stiffeners and girders and therefore depend upon the head to which they are subject.

Deep tank for oil fuel or oil cargo

A deep tank carrying oil will have a free surface, and, in the case of an oil fuel bunker, will have different levels of oil during the voyage. This results in reduced stability, while at the same time the momentum of the liquid moving across the tank may cause damage to the structure. To reduce this surging it is necessary to fit a centreline bulkhead if the tank extends from side to side of the ship. This bulkhead may be intact, in which case it must be as strong as the boundary bulkheads, or perforated, when the stiffeners may be considerably reduced.

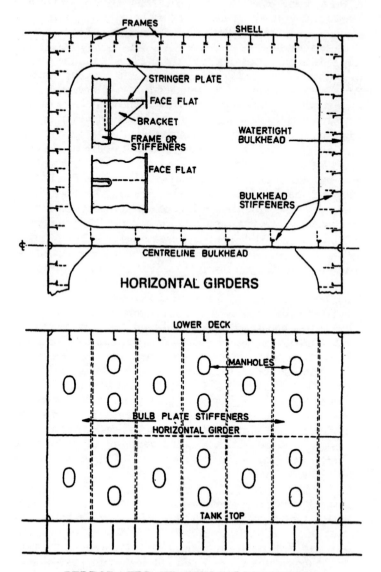

HORIZONTAL GIRDERS

PERFORATED CENTRELINE BULKHEAD

Fig. 6.7

The perforations must be between 5% and 10% of the area of the bulkhead. Any smaller area would allow a build-up of pressure on one side, for which the bulkhead is not designed, while a greater area would not reduce surging to any marked extent. Sparring must be fitted to the cargo side of a bulkhead which is a partition between a bunker and a hold. If a settling tank is heated and is adjacent to a compartment which may carry coal or cargo, the structure outside the tank must be insulated. Fig. 6.7 shows the structural arrangement of a deep tank which may be used for oil or dry cargo.

If dry cargo is to be carried in a deep tank, one or two large watertight hatches are required in the deck as described in Chapter 5.

NON-WATERTIGHT BULKHEADS

Any bulkhead which does not form part of a tank or part of the watertight subdivision of the ship may be non-watertight. Many of these bulkheads are fitted in a ship, forming engine casings and partitions in accommodation. 'Tween deck bulkheads fitted above the freeboard deck may be of non-watertight construction, while many ships are fitted with partial centreline bulkheads if grain is to be carried. Centreline bulkheads and many deck-house bulkheads act as pillars supporting beams and deck girders, in which case the stiffeners are designed to carry the load. The remaining bulkheads are lightly stiffened by angle bars or welded flats.

CORRUGATED BULKHEADS

A corrugated plate is stronger than a flat plate if subject to a bending moment or pillar load along the corrugations. This principle may be used in bulkhead construction, when the corrugations may be used to dispense with the stiffeners (Fig. 6.8), resulting in a considerable saving in weight. The troughs are vertical on transverse bulkheads but must be horizontal on continuous longitudinal bulkheads which form part of the longitudinal strength of the ship. A load acting across the corrugations will tend to cause the bulkheads to fold in concertina fashion. It is usual, therefore, on transverse bulkheads to fit a stiffened flat plate at the shell, thus increasing the transverse strength. This method also simplifies the fitting of the bulkhead to the shell which may prove difficult where the curvature of the shell is considerable. Horizontal diaphragm plates are fitted to prevent collapse of the troughs. These bulkheads form very smooth surfaces which, in oil tanks, allows improved drainage and ease of cleaning. A vertical stiffener is usually necessary if the bulkhead is required to support a deck girder.

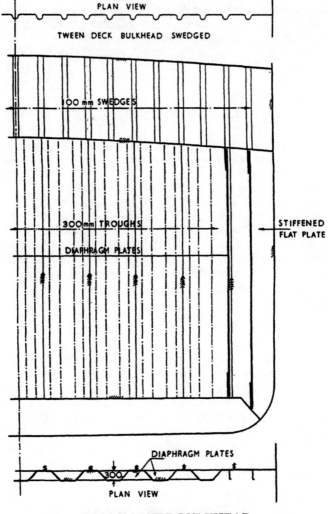

CORRUGATED BULKHEAD

Fig. 6.8

CHAPTER 7

FORE END ARRANGEMENTS

The structural arrangements of different ships vary considerably, depending largely upon whether the ship is riveted or welded. The more usual systems are given in this section.

STEM

The stem is formed by a solid bar which runs from the keel to the load waterline. In riveted ships this bar is rectangular, allowing the shell plating to be overlapped and riveted by two rows of rivets. The shell plating is stopped about 10 mm from the fore edge of the bar in order to protect the plate edges. At the bottom, the foremost keel plate is wrapped round the bar and, because of its shape, is known as a *coffin plate*. A similar form of construction is used at the top. In welded ships the bar is a solid round which improves the appearance considerably, particularly where the keel and side plates overlap.

Above the stem bar the stem is formed by plating which is strengthened by a welded stiffener on the centreline, the plating being thicker than the normal shell plating near the waterline but reduced in thickness towards the top. The plate stem is supported at intervals of about 1.5 m by horizontal plates known as *breast hooks,* which extend from the stem to the adjacent transverse frame. The breast hooks are welded to the stem plate and shell plating and are flanged on their free edge.

Modern stems are raked at 15° to 25° to the vertical, with a large curve at the bottom, running into the line of the keel. Above the waterline some stems curve forward of the normal rake line to form a *clipper* bow.

ARRANGEMENTS TO RESIST PANTING

The structure of the ship is strengthened to resist the effects of panting from 15% of the ship's length from forward to the stem and aft of the after peak bulkhead.

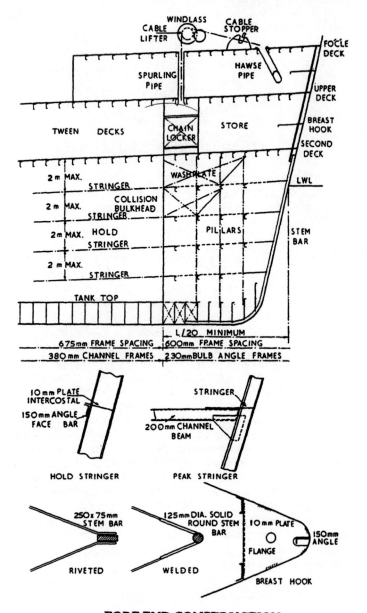

FORE END CONSTRUCTION

Fig. 7.1

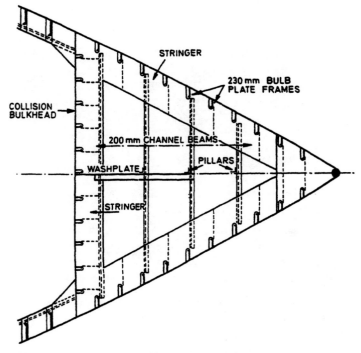

PANTING STRINGER

Fig. 7.2

In the fore peak, side stringers are fitted to the shell at intervals of 2 m below the lowest deck (Fig. 7.2). No edge stiffening is required as long as the stringer is connected to the shell, a welded connection being used in modern ships. The side stringers meet at the fore end, while in many ships a horizontal stringer is fitted to the collision bulkhead in line with each shell stringer. This forms a ring round the tank and supports the bulkhead stiffeners. Channel beams are fitted at alternate frames in line with the stringers, and connected to the frames by brackets. The intermediate frames are bracketed to the stringer. The free edge of the bulkhead stringer may be stiffened by one of the beams. In fine ships it is common practice to plate over the beams, lightening holes being punched in the plate.

The tank top is not carried into the peak, but solid floors are fitted at each frame. These floors are slightly thicker than those in the double bottom space and are flanged on their free edge.

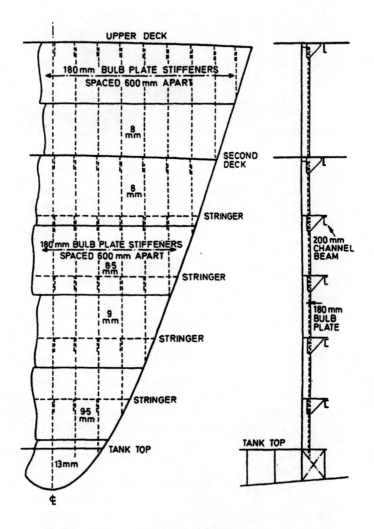

COLLISION BULKHEAD

Fig. 7.3

The side frames are spaced 610 mm apart and, being so well supported, are much smaller than the normal hold frames. The deck beams are supported by vertical angle pillars on alternate frames, which are connected to the panting beams and lapped onto the solid floors. A partial wash-plate is usually fitted to reduce the movement of the water in the tank. Intercostal plates are fitted for two or three frame spaces in line with the centre girder. The lower part of the peak is usually filled with cement to ensure efficient drainage of the space.

Between the collision bulkhead and 15% length from forward the main frames, together with their attachment to the margin, are increased in strength by 20%. In addition, the spacing of the frames from the collision bulkhead to 20% of the length from forward must be 700 mm. Light side stringers are fitted in the panting area in line with those in the peak. These stringers consist of intercostal plates connected to the shell and to a continuous face angle running along the toes of the frames. These stringers may be dispensed with if the shell plating is increased in thickness by 15%. This proves uneconomical when considering the weight but reduces the obstructions to cargo stowage in the hold. The peak is usually used as a tank and therefore such obstructions are of no importance.

The collision bulkhead is stiffened by vertical bulb plates spaced about 600 mm apart inside the peak. It is usual to fit horizontal plating because of the excessive taper on the plates which would occur with vertical plating. Fig. 7.3 shows the construction of a collision bulkhead.

The structure in the after peak is similar in principle to that in the fore peak, although the stringers and beams may be fitted 2.5 m apart. The floors should extend above the stern tube or the frames above the tube must be stiffened by flanged tie plates to reduce the possibility of vibration. The latter arrangement is shown in Fig. 8.8, chapter 8.

ARRANGEMENTS TO RESIST POUNDING

The structure is strengthened to resist the effects of pounding from the collision bulkhead to 25% of the ship's length from forward. The flat bottom shell plating adjacent to the keel on each side of the ship is increased in thickness by between 15% and 30% depending upon the length of the ship, larger ships having smaller increases.

In addition to increasing the plating, the unsupported panels of plating are reduced in size. In transversely framed ships the frame spacing in this region is 700 mm compared with 750 mm to 900 mm amidships. Longitudinal girders are fitted 2.2 m apart, extending vertically from the shell to the tank top, while intermediate half-height girders are fitted to the shell, reducing the unsupported width to 1.1 m. Solid floors are fitted at every frame space and are attached to the bottom shell by continuous welding.

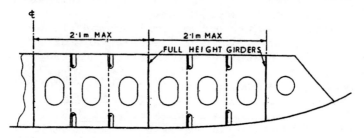

SOLID FLOOR IN POUNDING AREA

Fig. 7.4

If the bottom shell of a ship is longitudinally framed, the spacing of the longitudinals is reduced to 700 mm and they are continued as far forward as practicable to the collision bulkhead. The transverse floors may be fitted at alternate frames with this arrangement and the full-height side girders may be fitted 2.1 m apart. Half-height girders are not required (Fig. 7.4).

BULBOUS BOW

If a sphere is immersed just below the surface, and pulled through the water, a wave is created just behind the sphere. If such a sphere is superimposed on the stem of a ship, the wave from the sphere interferes with the normal bow wave and results in a smaller bow wave. Thus the force required to produce the bow wave is reduced. At the same time, however, the wetted surface area of the ship is increased, causing a slight increase in the frictional resistance. In slow ships the effect of a bulbous bow could be an increase in the total resistance, but in fast ships, where the wave making resistance forms a large proportion of the total resistance, the latter is reduced by fitting a bulbous bow. A bulbous bow also increases the buoyance forward and hence reduces the pitching of the ship to some small degree.

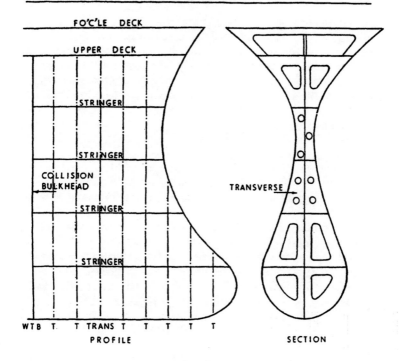

FO'C'LE DECK

UPPER DECK

STRINGER

STRINGER

COLLISION
BULKHEAD

STRINGER

STRINGER

TRANSVERSE

WTB T T TRANS T T T T T
PROFILE

SECTION

BULBOUS BOW

Fig. 7.5

The construction of the bulbous bow is shown in Fig. 7.5. The stem plating is formed by steel plates supported by a centreline web and horizontal diaphragm plates 1 m apart. The outer bulb plating is thicker than the normal shell plating, partly because of high water pressures and partly due to the possible damage by anchors and cables.

It is often found that due to the reduced width at the waterline caused by the bulb, horizontal stringers in the fore peak prove uneconomical and complete perforated flats are fitted.

ANCHOR AND CABLE ARRANGEMENTS

A typical arrangement for raising, lowering and stowing the anchors of a ship is shown in Fig. 7.1. The anchor is attached to a heavy chain cable which is led through the *hawse pipe* over the windlass and down through a *chain pipe* or *spurling pipe* into the chain locker.

The hawse pipes may be constructed of mild steel tubes with castings at the deck and shell, or cast in one complete unit for each side of the ship. There must be ample clearance for the anchor stock to prevent jamming and they must be strong enough to withstand the hammering which they receive from the cable and the anchor. The shell plating is increased in thickness in way of each hawse pipe and adjacent plate edges are fitted with mouldings to prevent damage. A chafing piece is fitted to the top of each hawse pipe, while a sliding cover is arranged to guard the opening.

The *cable stopper* is a casting with a hinged lever, which may be used to lock the cable in any desired position and thus relieve the load from the windlass either when the anchor is out or when it is stowed.

The drums of the windlass are shaped to suit the cable and are known as *cable lifters*. The cable lifters are arranged over the spurling pipes to ensure a direct lead for the cables into the lockers. The windlass may be either steam or electric in common with the other deck auxiliaries. Warping ends are fitted to assist in handling the mooring ropes. The windlass must rest on solid supports with pillars and runners in way of the holding down bolts, a 75 mm teak bed being fitted directly beneath the windlass.

The chain pipes are of mild steel, bell mouthed at the bottom. The bells may be of cast iron, well rounded to avoid chafing. The pipes are fitted as near as possible to the centre of the chain locker for ease of stowage.

The chain locker may be fitted between the upper and second decks, below the second deck or in the forecastle. It must be of sufficient volume to allow adequate headroom when the anchors are in the stowed position. The locker is usually situated forward of the collision bulkhead, using this bulkhead as the after locker bulkhead. The locker is not normally carried out to the ship side. The stiffeners are preferably fitted outside the locker to prevent damage from the chains. If the locker is fitted in the forecastle, the bulkheads may be used to support the windlass. A centreline division is fitted to separate the two chains and is carried above the stowed level of the chain but is not taken up to the deck. It is stiffened by means of solid half round bars while the top edge is protected by a split pipe. Foot holds are cut in to allow access from one side to the other. A hinged door is fitted in the forward bulkhead, giving access to the locker from the store space. Many lockers are fitted with false floors to allow drainage of water and mud, which is cleared by a drain plug in the forward bulkhead, leading into a drain hat from where it is discharged by means of a hand pump. The end of the cable must be connected to the deck or bulkhead in the chain locker. A typical arrangement is shown in Fig. 7.6.

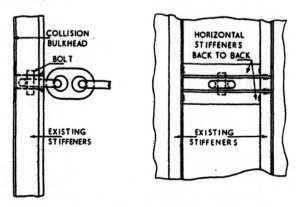

CONNECTION OF END OF CHAIN

Fig. 7.6

With this method, use is made of the existing stiffeners fitted to the fore side of the collision bulkhead. Two similar sections are fitted horizontally back to back, riveted to the bulkhead and welded to the adjacent stiffeners. A space is allowed between the horizontal bars to allow the end link of the cable to slide in and be secured by a bolt.

CHAPTER 8

AFTER END ARRANGEMENTS

CRUISER STERN

The cruiser stern forms a continuation of the hull of the ship above the sternframe and improves the appearance of the ship. It increases the buoyancy at the after end and improves the flow of water. Unfortunately it is very susceptible to slamming and must therefore be heavily stiffened. Lloyd's Rules require that the framing should be similar to that in the after peak with web frames fitted where necessary, while solid floors must be fitted, together with a centreline girder. In practice the structure is arranged in two main forms:

    (a) cant frames in conjunction with cant beams.

    (b) horizontal frames in conjunction with either cant beams or transverse beams.

A *cant frame* is one which is set at an angle to the centreline of the ship. Such frames are fitted 610 mm apart, thus dividing the perimeter of the cruiser stern into small panels. At the top, these frames are bracketed to *cant beams* which also lie at an angle to the centreline. The forward ends of the cant beams are connected to a deep beam extending right across the ship. At the lower ends, the cant frames are connected to a solid floor (Fig. 8.1).

The alternative method of construction has proved very successful, particularly with prefabricated structures. The horizontal frames are fitted at intervals of about 750 mm and are connected at their forward ends to a heavier transverse frame. They are supported at the centreline by a deep web which is also required with the cant frame system. If cant beams are fitted, the end brackets are carried down to the adjacent horizontal frame.

The structure at the fore end of the cruiser stern consists of solid floors attached to vertical side frames with transverse beams extending across the decks. A watertight rudder trunk is fitted enclosing part of the rudder stock. A typical centreline view of a cruiser stern with horizontal framing is shown in Fig. 8.2.

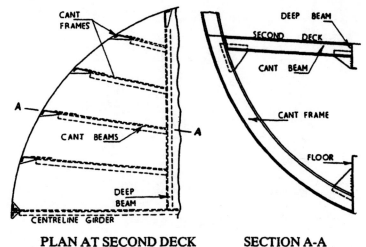

PLAN AT SECOND DECK   SECTION A-A

Fig. 8.1

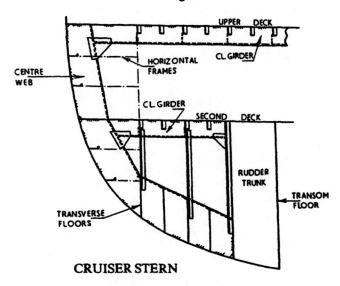

CRUISER STERN

Fig. 8.2

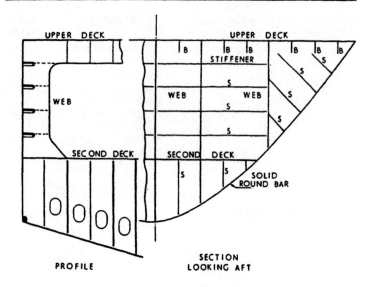

TRANSOM STERN

Fig. 8.3

In many larger ships a transom stern is fitted, *i.e.,* the stern is flat (Fig. 8.3). This form of stern reduces the production costs while at the same time reducing the bending moment on the after structure caused by the unsupported overhang. The stiffening is invariably horizontal.

STERNFRAME AND RUDDER

The sternframe forms the termination of the lower part of the shell at the after end of the ship. In single screw ships the sternframe carries the boss and supports the after end of the sterntube. The rudder is usually supported by a vertical post which forms part of the sternframe. It is essential, therefore, that the structure should be soundly constructed and of tremendous strength. Sternframes may be cast, fabricated or forged, the latter method of construction having lost its popularity although parts of the fabricated sternframe may be forged. Both fabricated and cast sternframes may be shaped to suit the form of the hull and streamlined to reduce turbulence of the water. The choice of casting or fabrication in the construction of a sternframe depends upon the personal preference of the shipowner and shipbuilder, neither method having much advantage over the other.

It is useful to consider the rudder in conjunction with the sternframe. There are no regulations for the area of a rudder, but it has been found in practice that an area of between one sixtieth and one seventieth of the product of the length and draught of the ship provides ample manoeuvrability for deep sea vessels. The ratio of depth to width, known as the aspect ratio, is usually about two. Single plate rudders were used for many years but are now seldom used because of the increased turbulence they create. Modern rudders are streamlined to reduce eddy resistance. If part of the rudder area lies forward of the turning axis, the turning moment is reduced and hence a smaller rudder stock may be fitted. A rudder with the whole of its area aft of the stock is said to be *unbalanced*. A rudder with between 20% and 40% of its area forward of the stock is said to be *balanced,* since at some rudder angle there will be no torque on the stock. A rudder which has part of its area forward of the stock, but at no rudder angle is balanced, is said to be *semibalanced.*

Fabricated sternframe with unbalanced rudder

Fig. 8.4 shows a fabricated sternframe used to support an unbalanced rudder. The *sole piece* is a forging which is carried aft to form the *lower gudgeon* supporting the bearing pintle, and forward to scarph to the aftermost keel plate which is known as a *coffin plate* because of its shape. The sternpost is formed by a solid round bar to which heavy plates, 25 mm to 40 mm thick, are welded, the boss being positioned to suit the height of the shaft. Thick web plates are fitted horizontally to tie the two sides of the sternframe rigidly. The side shell plates are riveted or welded to the plates forming the sternframe. This form of construction is continued to form the *arch* which joins the sternpost to the rudderpost. Vertical webs are used in this position to secure the sternframe to the floor plates, while a thick centreline web is fitted to ensure rigidity of the arch.

The rudderpost is similar in construction to the sternpost, a thick web plate being fitted at the after side, while one or more gudgeons are fitted as required to support the rudder. The web plate is continued inside the ship at the top of the rudder post and attached to a thick transom floor which is watertight.

The rudder is formed by two plates about 10 mm to 20 mm thick, connected at the top and bottom to forgings which are extended to form the upper and lower gudgeons. The upper

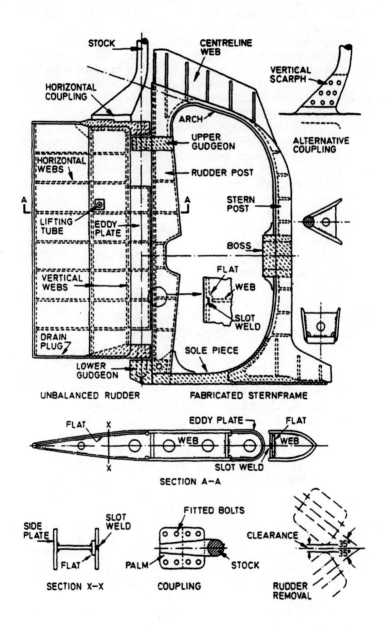

Fig. 8.4

forging is opened into a palm, forming part of the horizontal coupling. This palm is stepped to provide a shoulder which reduces the possibility of shearing the bolts. The side plates are stiffened by means of vertical and horizontal webs. If the structure is riveted, tap rivets must be used on one side since the points of normal rivets would be inaccessible. The structure is difficult to weld for the same reason. An efficient attachment may be made by fitting a flat bar to the edge of the horizontal webs and slot welding. One disadvantage of double plate rudders is the possibility of internal corrosion. The inner surfaces must be adequately protected by some form of coating, while a drain plug must be fitted to avoid accumulations of water.

The rudder is supported by *pintles* which fit into the gudgeons, Fig. 8.5. The upper part of each pintle is tapered and fits into a similar taper in the rudder gudgeons. The pintle is pulled hard against the taper by means of a large nut with some suitable locking device, such as a lock nut or split pin. A brass liner is fitted round the lower part of the pintle. Lignum vitae or laminated plastic is dovetailed into the sternframe gudgeon to provide a bearing surface for the pintle, allowing the pintle to turn but preventing any side movement. A head is fitted to the upper pintle to prevent undue vertical movement of the rudder. This is known as a *locking pintle*. The bottom pintle is known as a *bearing pintle* since it rests on a hardened steel pad shaped to suit the bottom of the pintle. A hole is drilled in the gudgeon,

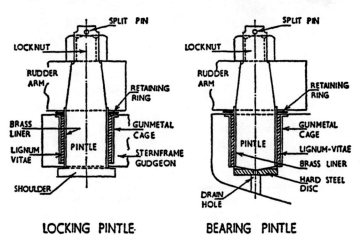

LOCKING PINTLE·   BEARING PINTLE

Fig. 8.5

with a smaller hole in the bearing pad, to allow for the free circulation of water which acts as a lubricant for the lignum vitae, and allows the bearing pad to be punched out when worn.

The rudder is turned by means of a stock which is of forged steel, opened out into a palm at its lower end. The stock is carried through the rudder trunk and keyed to the steering engine. *It is essential that the centreline of stock and centreline of pintles are in the same line,* otherwise the rudder will not turn. A watertight gland must be fitted round the stock where it penetrates the deck. Many ships, however, are fitted with rudder carriers (Fig. 8.6), which themselves form watertight glands. The bearing surfaces are formed by cast iron cones, the upper cone being fitted to the rudder stock. As the bearing surfaces of the

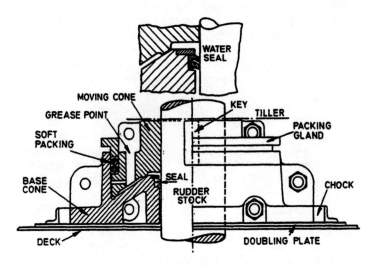

RUDDER CARRIER
Fig. 8.6

lower pintle wear, the weight of the rudder will be taken by the carrier, and therefore the vertical wear down should be very small. Indeed, it is found in practice that any appreciable wear down is the result of a fault in the bearing surfaces, usually due to the misalignment of the stock. This causes uneven wasting of the surface and necessitates refacing the bearing surfaces and re-aligning the stock. In most cases, however, the cast iron work hardens and forms a very efficient bearing surface.

To remove the rudder it is first necessary to remove the locking pintle. The bearing may not be removed at this stage.

The rudder is turned by means of the stock to its maximum angle of, say, 35° on one side. The bolts in the coupling are removed and the stock raised sufficient to clear the shoulder on the palm. The stock is turned to the maximum angle on the opposite side, when the two parts of the coupling must be clear. The rudder may then be removed or the stock drawn from the ship.

Cast steel sternframe with balanced rudder

A cast steel sternframe is shown in Fig. 8.7. The casting may be in one or two pieces, the latter reducing the cost of repair in the event of damage. The sole piece is carried forward and scarphed to the aftermost keelplate, while the after end forms the lower gudgeon. The sternpost is carried up inside the ship and opened to form a palm which is connected to a floor plate. This is known is a vibration post. The casting is continued aft at the top of the propeller aperture to form the arch and the upper rudder support. At the extreme after end the casting is carried inside the ship and opened into a palm which is connected to a watertight transom floor. The two parts of the casting are joined together by a combination of riveting and welding. The two webs forming the ends of the separate parts are riveted together, while the joint on the outside of the casting is welded.

The rudder is constructed of double plates, with a large tube down the centre. The rudder post is formed by a detachable forged steel mainpiece which is carried through the tube, bolted to a palm on the stern frame at the top and pulled against a taper in the lower gudgeon. The mainpiece is increased in diameter at the top and bottom where lignum vitae bearing strips are fitted. Castings are fitted at the top and bottom of the tube to carry the bearing strips. Hard steel bearing rings are fitted between the rudder and the bottom gudgeon to take the weight of the rudder. A horizontal coupling is shown in the diagram, attaching the stock to the rudder with the aid of fitted bolts. There must be sufficient vertical clearance between the stock and the mainpiece to allow the mainpiece to be raised sufficiently to clear the bottom gudgeon when removing the rudder. The upper stock is usually supported by a rudder carrier. By balancing a rudder in a particular ship, the diameter of the stock was reduced from 460 mm to 320 mm. This allows reduction in the thickness of the side plates and the size of the steering gear.

A flat bar is welded to the bottom of the horn to restrict the lift of the rudder. The clearance between the rudder and the flat should be less than the cross-head clearance. Any vertical force on the rudder will hence be transmitted to the sternframe and not to the steering gear.

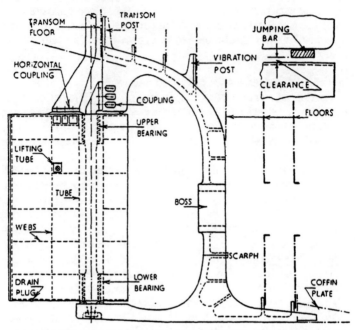

BALANCED RUDDER   CAST STEEL STERNFRAME

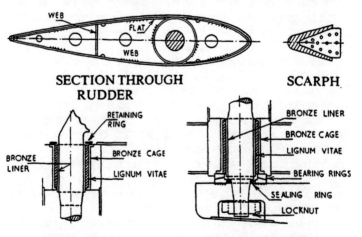

SECTION THROUGH
RUDDER   SCARPH

UPPER BEARING   Fig. 8.7   LOWER BEARING

Open water stern with spade rudder

Many modern large ships are fitted with spade-type rudders (Fig. 8.8). The rudder is supported by means of a gudgeon on a large rudder horn and by the lower end of the stock, the latter being carried straight into the rudder and keyed.

The lower part of the ship at the after end is known as the *deadwood* since it serves no useful purpose. The spade rudder is unsupported at the bottom and hence an open aperture is possible. This allows the deadwood to be cut away, resulting in a better flow of water to the propeller. In addition, the distance of the rudder from the propeller may be adjusted to improve the efficiency of the rudder and in practice considerable reductions have been made, both in the diameter of the turning circle and in the vibration when turning.

The rudder horn and the sternframe may be cast or fabricated.

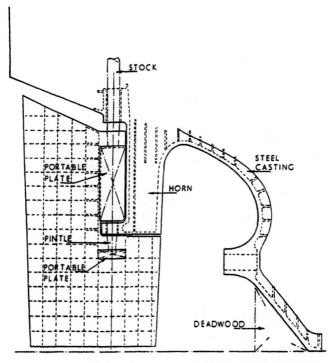

SPADE RUDDER   &   OPEN WATER STERN

Fig. 8.8

Rudder and sternframe for twin screw ship

In a twin screw ship the propellers are fitted off the centreline of the ship and therefore no aperture is required in the sternframe which may then be designed only to support the rudder. In a single screw ship the deadwood is used to assist in the support of the sole piece but this is not necessary in twin screw vessels. Many designers make use of this space by carrying the lower part of the rudder forward of the centreline of stock. Fig. 8.9 shows a typical arrangement.

The sternframe is of cast steel, constructed to cut out the deadwood, and notched or rebated to suit the shell plating. The top of the casting is connected by means of a palm to the transom floor.

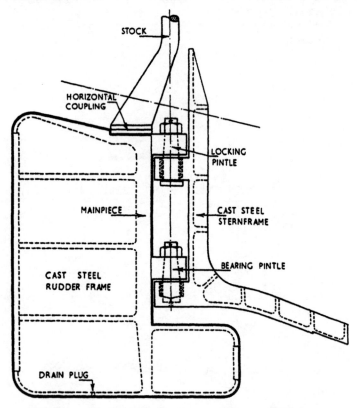

SEMI-BALANCED RUDDER — TWIN SCREW SHIP

Fig. 8.9

The rudder shown in the diagram has a frame of cast steel, the frame consisting of a mainpiece with horizontal arms which are of streamlined form. The side plates are riveted or welded to the frame. There must be sufficient clearance between the lower part of the sternframe and the extension of the rudder to allow the rudder to be lifted clear of the bearing pintle. The mainpiece must be particularly strong to prevent undue deflection of the lower, unsupported portion.

Bossings and spectacle frame for twin screw ship

The shafts of a twin screw ship are set at a small angle to the centreline. As the width of the ship reduces towards the after end, the shaft projects through the normal line of the shell. The shell widens out round the shaft to form the *bossings* which allow access to the shaft from inside the ship and allows bearings

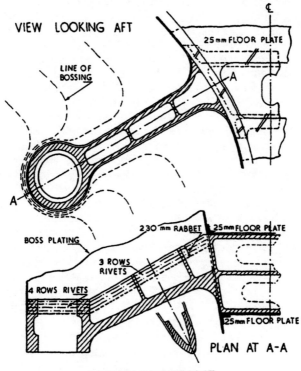

SPECTACLE FRAME

Fig. 8.10

to be fitted where required. The after end of the bossing terminates in a casting which carries the boss and supports the after end of the stern tube. This casting must be strongly constructed and efficiently attached to the main hull structure to reduce vibration. It is usual to carry the casting in one piece across the ship, thus forming the *spectacle frame* (Fig. 8.10). The centre of the casting is in the form of a box extending over two frame spaces and attached to thick floor plates.

SHAFT TUNNEL

When the machinery space is divided from the after peak by one or more cargo holds, the main shafting must be carried through the holds. A tunnel is then built round the shaft to prevent contact with the cargo and to give access to the shaft at all times for maintenance, inspection and repair. The tunnel is watertight and extends from the after machinery space bulkhead to the after peak bulkhead. It is not necessary to provide a passage on both sides of the shaft, and the tunnel is therefore built off the centreline of the ship, allowing a passage down the starboard side. The top of the tunnel is usually circular except in a deep tank when it is more convenient to fit a flat top. Fig. 8.11 shows a cross-section through a shaft tunnel.

The tunnel stiffeners or *rings* are fitted inside the tunnel although in insulated ships and in tunnels which pass through deep tanks, the rings are fitted outside the tunnel. The rings may be welded to the tank top or connected by angle lugs. The plating is attached to the tank top by welding or by a boundary angle fitted on the opposite side of the plating to the stiffeners. The stiffeners and plating must be strong enough to withstand a water pressure without appreciable leakage in the event of flooding. The scantlings are therefore equivalent to those required for watertight bulkheads. Under the hatches the tunnel top plating is increased by 2 mm unless wood sheathing is fitted. One of the side plates is arranged so that it may easily be removed, together with the stiffeners, to allow the main shafting to be unshipped.

A watertight door is fitted in the machinery space bulkhead giving access to the shaft tunnel from the machinery space (see Chapter 6). At the after end of the tunnel, a watertight escape trunk is fitted and extends to the deck above the load waterline. At the after end of the tunnel, the ship is so fine that there is very little useful cargo space at each side of the tunnel. The tunnel top is then carried right across the ship to form a *tunnel recess*. The

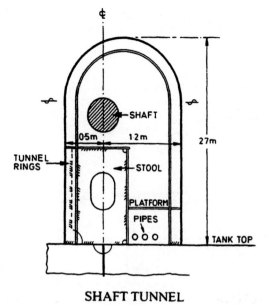

SHAFT TUNNEL

Fig. 8.11

additional space on the port side of this recess is usually used to store the spare tail shaft.

The shaft tunnel is used as a pipe tunnel, the pipes being carried along the tank top with a light metal walking platform fitted about 0.5 m from the tank top. The shaft is supported at intervals by bearings which are fitted on shaft stools. The tops of the stools are lined up accurately to suit the height of the shaft, although adjustments to the height of bearings are made when the ship is afloat. The stools are constructed of 12 mm plates, riveted or welded together, the latter being the most usual. They are attached to the tunnel rings to prevent movement of the bearings which could lead to damage of the shaft. The loads from the bearings are transmitted to the double bottom structure by means of longitudinal brackets. Manholes are cut in the end plates to reduce the weight and to allow inspection and maintenance of the stools.

KORT NOZZLE

The Kort Nozzle is a form of hollow truncated cone which is fitted around the propeller in order to increase the propulsive efficiency. Two types of nozzle are available, the fixed nozzle which is welded to the ship and forms part of the hull structure, and the nozzle rudder which replaces the normal rudder.

Fixed nozzle

Fig. 8.12 shows the arrangement of a nozzle which is fixed in joined together by plates which form aerofoil cross-section. At the bottom the nozzle is welded to the sole piece of the sternframe, while at the top it is faired into the shell plating. Diaphragm plates are fitted at intervals to support the structure.

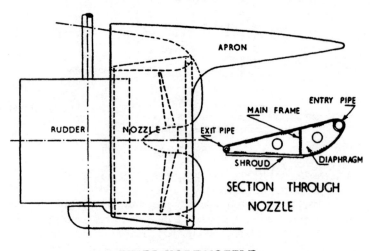

FIXED KORT NOZZLE

Fig. 8.12

The nozzle directs the water into the propeller disc in lines parallel to the shaft, causing an increase in thrust of 20% to 50%. This may be used in several ways. The ship speed may be increased by 5% to 10% with no increase in power. The power may be reduced for the same speed or, in vessels such as tugs, the increase in thrust or towing force may be accepted for the same power. It is also found that, in rough water, the effect of pitching on the propulsive efficiency is greatly reduced, since the water is still directed into the propeller disc.

In all rotating machinery the clearance between the rotating element and the casing should preferably be constant, maximum efficiency being obtained when the clearance is in the order of one thousandth of the diameter. The Kort Nozzle permits a constant clearance, although this is in the order of one hundredth of the diameter. One of the troubles with conventional type sternframes is the fluctuation in stress on the propeller blades as they pass close to the structure. If the tip clearance is too small, these fluctuations cause vibration of the propeller blades. Such vibration is avoided entirely with the constant tip clearance of the Kort Nozzle.

The nozzle also acts as a guard for the propeller, protecting it from loose ropes and floating debris. If, however, ropes do become entangled with the propeller, they are much more difficult to remove than with the open propeller.

Such nozzles are fitted mainly to tugs and trawlers, where the increased pull may be utilised directly in service.

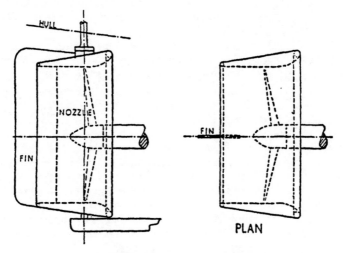

KORT NOZZLE RUDDER

Fig. 8.13

Nozzle rudder

One of the disadvantages of the fixed nozzle was the difficulty of moving astern, the after end tending to drift. This is overcome by fitting a centreline fin plate to the after end of a nozzle which turns about a vertical stock and thus dispenses with the normal form of rudder (Fig. 8.13). The water is then projected at an angle to the centreline, causing the ship to turn. The centreline of the stock must be in line with the propeller in order to allow the nozzle to turn and yet maintain small tip clearance.

The support of the shaft at the extreme after end may prove rather difficult with this form of construction. The bossing may be increased and extended, or the boss may be supported by brackets fitted to the stern.

Both types of nozzle may be fitted to existing ships, although the rudder type requires greater alteration to the structure.

TAIL FLAPS AND ROTATING CYLINDERS

It was stated earlier in this chapter that the rudder angle is usually limited to about 35° on each side of the centreline. It is found that if this angle is exceeded the diameter of the turning circle is increased, largely due to the separation of the flow of water behind the rudder. In vessels where high manoeuvrability is essential this limit is a disadvantage.

One method of improving the rudder performance is to fit a tail flap which moves automatically to a larger angle as the rudder is turned, in a similar manner to a fin stabiliser (see Chapter 11). Experiments have shown excellent results although the cost of manufacture and maintenance would preclude the fitting of such a device in a normal ship.

An alternative method which has been tested is to fit a rotating cylinder at the fore end of the rudder. This cylinder controls the boundary layer of water and reduces the separation of the water behind the rudder, producing a positive thrust at larger rudder angles with consequent large reductions in diameter of turning circle and increased rate of turn. Typically a large tanker travelling at 15 knots would have a cylinder of 1 m diameter rotating at about 350 rev/min and requiring about 400 kW. Although positive thrust is achieved at a rudder angle of 90°, practical considerations would limit the rudder angle to about 70° and even this would require a re-design of steering gears.

CHAPTER 9

OIL TANKERS, BULK CARRIERS, LIQUEFIED GAS CARRIERS AND CONTAINER SHIPS

OIL TANKERS

There has been a tremendous growth in the size of tankers since the end of World War II. The deadweight of these vessels has increased from about 13 000 tonne in 1946 to the present V.L.C.C.s *(very large crude carriers)* of 150 000 tonne to 250 000 tonne and the U.L.C.C.s *(ultra large crude carriers)* of over 300 000 tonne.

The design of the structure has progressed in much the same way. Many tankers built in 1946 were of riveted construction, while the construction of the welded ships left much to be desired. The majority of the faults in these tankers have been overcome in modern vessels by using continuous, welded structure with well-rounded corners. At the same time improved quality steel has been introduced which is less susceptible to the formation of cracks and some, known as *extra-notch tough steel,* which will prevent the spread of cracks.

The structural arrangements and details vary considerably from shipyard to shipyard, but there are two basic methods of framing in use:

(i) longitudinal framing in which the deck, bottom shell, side shell and longitudinal bulkheads are stiffened longitudinally (Fig. 9.1).

(ii) combined framing in which the deck and bottom shell are framed longitudinally, with transverse side frames and vertical stiffeners on the longitudinal bulkheads (Fig. 9.2).

Longitudinal framing

The longitudinal framing system provides ample longitudinal strength, but the horizontal side frames and longitudinal bulkhead stiffeners are likely to retain liquid and hence increase the corrosion. The longitudinals are supported by deep transverse webs which form rings round the ship at intervals of 3 m to 6 m and to which they are attached by flat bars or brackets. At the transverse bulkheads, the structure must be carefully designed to give maximum continuity of strength. Fig. 9.3 shows typical attachments of bottom and side longitudinals at the transverse bulkhead.

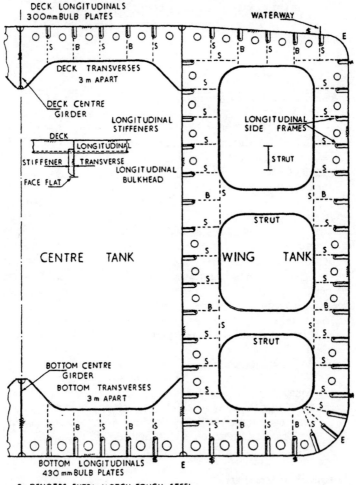

OIL TANKER   LONGITUDINAL FRAMING   Fig. 9.1

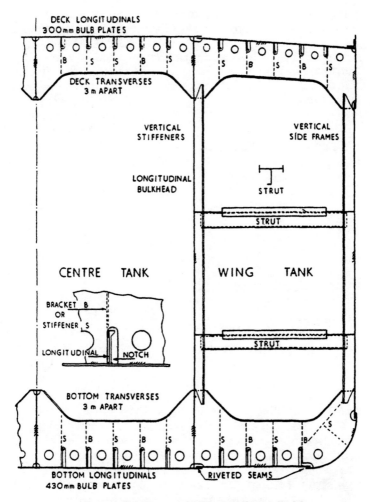

OIL TANKER   COMBINED FRAMING

Fig. 9.2

The longitudinals are usually bulb plates although many of the larger vessels employ large flat plates. The bottom longitudinals are much heavier than the deck longitudinals, while the side longitudinals increase with the depth of the tank. The transverse webs fitted to the side shell and longitudinal bulkhead are strengthened by face flats and supported by two or three horizontal struts arranged in such a way that the unsupported span of transverse at the top is greater than that at the bottom, the latter being subject to a greater head of liquid. It is intended by this form of design, that the bending moments on the separate spans should be equal. It is essential that the face flat on the web is carried round the strut to form a continuous ring of material.

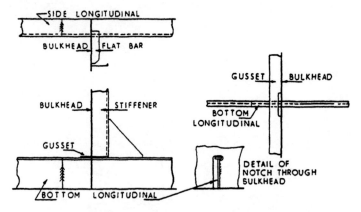

END CONNECTIONS OF LONGITUDINALS

Fig. 9.3

Combined framing

This system has proved successful in many ships, having the advantage of providing sufficient longitudinal strength with good tank drainage due to the vertical side frames and stiffeners. The latter are supported by horizontal stringers which are continuous between the transverse bulkheads and tied at intervals by struts. The lower stringers are heavier than the upper stringers. Where the length of the tanker exceeds 200 m Lloyd's require longitudinal framing to be used.

General

A deep centreline girder must be fitted at the keel and deck, connected to a vertical web on the transverse bulkhead. This web is only required on one side of the bulkhead. The top and bottom girders act as supports for the transverses and hence reduce their span. Large face flats are fitted to their free edges and continued round the bulkhead web to form almost a complete vertical ring (Fig. 9.4). The bottom centre girder is required to support the ship while in dock and it is found necessary for this reason to reduce the unsupported panels of keel plate by fitting docking brackets between the transverses, extending from the centre girder to the adjacent longitudinal (see Chapter 2). In many of the larger ships the centre girders have been replaced by a full-height perforated bulkhead.

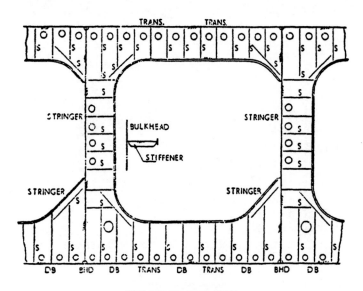

CENTRELINE WEB

Fig. 9.4

The transverse bulkheads are usually stiffened vertically, the stiffeners being bracketed at their ends and supported by horizontal stringers (Fig. 9.5). Corrugated bulkheads are often fitted, the corrugations being vertical, and have the advantages of improving drainage, allowing easy cleaning and reducing weight. The longitudinal bulkheads may also be corrugated but in this case the corrugations must be horizontal, otherwise the longitudinal strength would be impaired as the bulkhead would tend to fold like a concertina as the ship hogs and sags.

The thickness of the deck plating depends on the maximum bending moment to which the ship is liable to be subject, and is given in the form of a cross-sectional area of material in way of openings. These openings consist of oiltight hatches and tank cleaning holes. When a bridge is fitted amidships it is often necessary to fit the hatches to all three tanks in the same transverse line. This results in extremely thick deck plating forward and aft of the bridge and is one of the many reasons for the transfer of the accommodation to the after end in some modern vessels. Fig. 9.6 shows a typical oiltight hatch which is fitted to many tankers. It is simple to open and close, with no danger of jammed fingers and may be handled by one man.

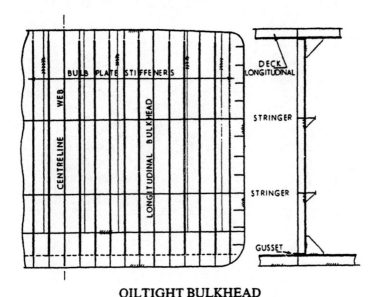

OILTIGHT BULKHEAD

Fig. 9.5

Throughout the ship the greatest care is taken to ensure continuity of the structure. At the ends, the longitudinals reduce in number gradually until, in the engine room and at the fore end, the ship is transversely framed. The ends of the longitudinal bulkheads are continued in the form of brackets, while in some designs the bulkheads are carried through the whole length of the machinery space, forming tanks, stores and workshop spaces at the sides.

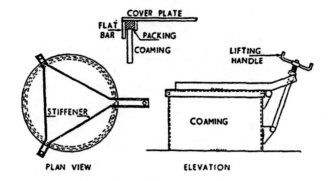

EASIFIT OILTIGHT HATCH

Fig. 9.6

Cargo pumping and piping arrangements

While it is not the intention in this book to deal with piping, it is perhaps necessary to include the method of loading and discharging oil cargoes.

A cargo pump room is arranged at the after end of the cargo tank range, containing three or four large capacity centrifugal pumps together with between two and four smaller capacity stripping pumps. The latter are used to clear the tanks of oil when the main cargo pumps lose suction. In addition, two ballast pumps may be provided.

Several systems of piping are in use, depending largely on whether it is intended to carry single grade or multi-grade cargo. A simple piping arrangement may be used for single grade cargo as shown in Fig. 9.7. This is known as a ring main system and consists of a continuous pipe from the pump room to the forward cargo tank and back into the pump room, with cross-over pipes in each centre tank, extending into the wing tanks. This system may be used in conjunction with two or three pumps, the centre pump assisting one or both of the outside pumps. It is possible to carry two different grades of cargo with this system and to discharge them both at the same time. Valves with extended spindles up to the deck are fitted in each tank, allowing the tank to be discharged or by-passed as required. The oil is discharged through pipes fitted on the upper deck amidships and led aft along the deck to the pump room, while a single stern discharge is usually arranged.

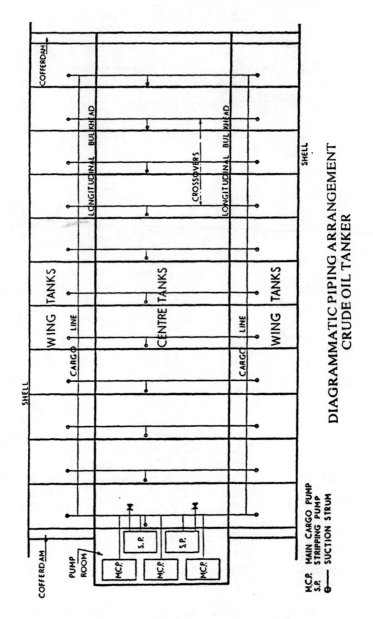

DIAGRAMMATIC PIPING ARRANGEMENT
CRUDE OIL TANKER

M.C.P.   MAIN CARGO PUMP
S.P.   STRIPPING PUMP
⊖———   SUCTION STRUM

Fig. 9.7

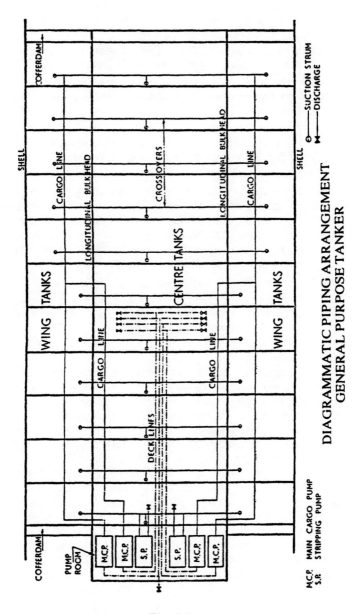

**DIAGRAMMATIC PIPING ARRANGEMENT
GENERAL PURPOSE TANKER**

M.C.P.   MAIN CARGO PUMP
S.P.   STRIPPING PUMP

Fig. 9.8

A more complicated piping arrangement is required when several grades of cargo are carried and one such arrangement is shown in Fig. 9.8. In this case four main cargo pumps may be used to discharge the separate cargoes which are drawn from the tanks through individual lines. Two smaller capacity stripping pumps are fitted with connections to the main cargo tank lines.

A modern arrangement has been developed in which the oil is allowed to flow through hydraulically-operated sluice valves into a common suction strum from where it is discharged by the main cargo pumps. This system may be used for single grade cargo and is intended to dispense with the cargo pipelines in the tanks. In case of failure of the valves, however, a heavy duty stripping pump and line is fitted.

Many oils must be heated before discharging and therefore heating coils are fitted to all tanks. The coils are of cast iron with large heating surfaces which are heated by means of steam. It is essential that these coils should be close to the bottom of the tank and it is therefore necessary to cut holes in the bottom transverses to allow the coils to be fitted.

While many different types of pump are available, several ships make use of vertical drive pumps in which the driving spindles extend through an oiltight flat into the machinery space, the motors being fitted on the flat. This ensures regular maintenance by the engine room staff without entering the pump room.

When cargo is being loaded or discharged using the high capacity pumps, the structure is subject to an increase or reduction in pressure. Similar variations in pressure occur due to the thermal expansion or contraction of the cargo. To avoid structural damage, some form of pressure/vacuum release must be fitted to accommodate these variations. In addition, provision must be made to ventilate the gas-filled space, ensuring that the gas is ejected well above the deck level. For many years the latter was achieved by running vapour lines from the hatches to a common vent pipe running up the mast and fitted with a flame arrester at the top. More recently individual high velocity vents on stand pipes have been fitted to the hatches. By restricting the orifice, a high gas velocity is produced when loading. A pressure/vacuum valve is incorporated in the system. Fig. 9.9 shows a typical system in the open position.

In order to reduce the possibility of explosion in the cargo tanks the oxygen content is reduced by means of an inert gas system. It is essential that the void space in the tanks is inerted at all times, *i.e.* when full, when empty and particularly during the tank cleaning operation.

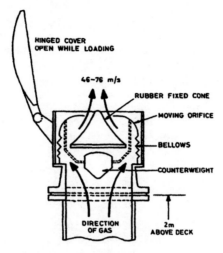

HINGED COVER
OPEN WHILE LOADING

46-76 m/s

RUBBER FIXED CONE

MOVING ORIFICE

BELLOWS

COUNTERWEIGHT

DIRECTION
OF GAS

2m
ABOVE DECK

CARGO GAS VENTING SYSTEM   Fig. 9.9

Crude oil washing

Once cargo has been discharged it is necessary to clean the tanks of the sludge which accumulates on horizontal surfaces and in the bottom of the tanks. Until recently the tanks were water washed using high pressure nozzles from either portable or fixed machines. Even when using hot water, however, not all the sludge was removed and it was necessary at times to dig out the heavy deposits. In addition, it is suspected that water washing contributed to the cause of explosions on three VLCC.

Crude oil is itself a solvent and is now used as a washing agent as an alternative to water in crude oil carriers. The crude oil washing (COW) system comprises a permanent arrangement of steel piping, independent of the fire mains or any other piping system, connected to high capacity washing machines having nozzles which rotate through 120° and project the crude oil at about 9 to 10 bar. The supply to the machines is provided either by the main cargo pumps or by pumps provided for that purpose.

The machines are so located that all horizontal and vertical areas within the tank are washed either by direct impingement or by deflection or splashing of the jet. The number and disposition of the machines will depend upon the structural arrangements within the tank, but less than 10% of the total surface of the structure may be shielded from direct impingement.

It is essential that an efficient inert gas system is used in conjunction with COW. It is also a requirement that only trained personnel carry out the work. A comprehensive operations and equipment manual is required on each ship.

The advantages claimed for COW are:
(i) Less sea pollution due to reduced water washing.
(ii) Reduction in time and cost of tank cleaning.
(iii) No de-sludging by hand required.
(iv) Increased cargo deadweight due to the removal of the sludge and less oil/water slops retained on board.
(v) Reduced tank corrosion caused by water washing.

The disadvantages associated with the system are:
(i) Increase in discharge time.
(ii) Increased work load on ship staff.

BULK CARRIERS

During the past few years a large proportion of bulk carriers have been built. While their increase in size has not been of the same order as oil tankers, the average deadweight of these ships is still high. The majority of bulk cargoes were, until recently, carried by cargo tramps which were not designed specifically for such cargoes. The rate of loading in bulk at some terminals is tremendous, 4000 to 5000 tonne/hour being regarded as normal. When carrying a heavy cargo such as iron ore in a cargo tramp, it is advisable to load a proportion of the cargo in the 'tween decks, otherwise the vessel becomes too stable and hence not only uncomfortable but dangerous. This means that the rate of loading is considerably reduced. For this reason most cargo tramp owners have branched out into the specialised bulk carrier trade.

The design of a bulk carrier depends largely on whether it is intended to carry a particular cargo or whether any type of bulk cargo is to be catered for. A vessel required to carry iron ore has a small hold capacity since the ore is heavy. A deep double bottom is fitted, together with longitudinal bulkheads which restrict the ore and maintain a high centre of gravity consistent with comfortable rolling. A ship designed to carry bauxite, however, requires twice the volume of space for cargo and will therefore have a normal height of double bottom although longitudinal bulkheads may be used to restrict the ore space.

A bulk 'tramp' if one can coin a term, *i.e.,* one which may be required to carry any type of bulk cargo, must have restricted volume for an iron ore cargo and at the same time must have sufficient cargo capacity to carry its full deadweight of light grain which requires three times the volume of the ore. One method of overcoming this difficulty is to design the ship to load ore in alternate holds.

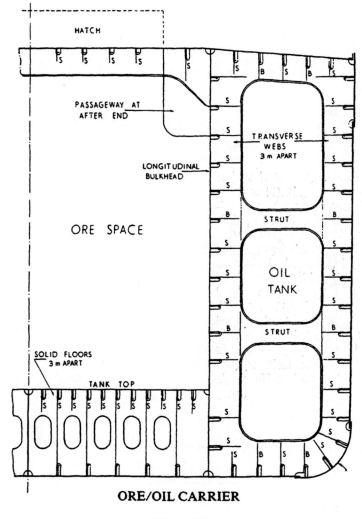

ORE/OIL CARRIER

Fig. 9.10

It may readily be seen, therefore, that the design of bulk carriers will vary considerably. Fig. 9.10 shows a cross-section of a bulk ore carrier (OBO) which may carry an alternative cargo of oil in the wings and double bottom.

The structure is similar to that required for oil tankers, having longitudinal framing at the deck, bottom and side shell, longitudinal bulkhead and tank top. These longitudinals are supported by transverse webs 2.5 m apart. The supporting

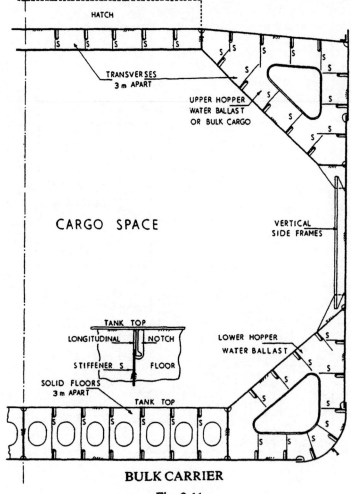

HATCH

TRANSVERSES
3 m APART

UPPER HOPPER
WATER BALLAST
OR BULK CARGO

CARGO SPACE

VERTICAL
SIDE FRAMES

TANK TOP

LONGITUDINAL   NOTCH

STIFFENER S   FLOOR

SOLID FLOORS
3 m APART

TANK TOP

LOWER HOPPER
WATER BALLAST

BULK CARRIER

Fig. 9.11

members are, as far as possible, fitted in the tanks rather than the ore space, to facilitate discharge of cargo using grabs. For the same reason it is common practice to increase the thickness of the tank top beyond that required by the Classification Societies.

The structural arrangement of a bulk carrier is shown in Fig. 9.11. The arrangement of the stiffening members is once again similar to the oil tanker, although the layout of the cargo space is entirely different from that shown in Fig. 9.10. In this design the double bottom, lower hopper and upper hopper spaces are available for water ballast, the upper tanks raising the centre of gravity of the ship and hence reducing the stiffness of roll. A large cargo capacity is provided by the main cargo holds and, if necessary, the upper hopper tanks.

LIQUEFIED GAS CARRIERS

For economical reasons it is necessary to carry gas in liquefied form rather than as a vapour, the volume of the liquid being $\frac{1}{300}$ to $\frac{1}{600}$ of the volume of the vapour. Several gases are transported in this way, such as methane, propane, butane and anhydrous ammonia. The gases are divided into *liquefied natural gas* (LNG) which consists mainly of methane and *liquefied petroleum gas* (LPG), mainly propane, propylene, butane and butylene. The latter are derived from the refining of the LNG or as a by-product of the distillation of crude oil at the refineries.

Methane has a boiling point at atmospheric pressure of $-162°C$, whilst its critical temperature is $-82°C$ at a pressure of 47 bar, *i.e.*, the gas cannot exist as a liquid at a temperature higher than $-82°C$, no matter what the pressure is. Thus the containment system for LNG must be suitable for conditions between these limits. It is usually more economical to design for the lower temperature at atmospheric pressure.

LPG requirements vary between maximum required pressures of about 18 bar to atmospheric pressure, and minimum temperatures of about $-45°C$ (ethylene $-104°C$) to $-5°C$.

Many different types of tank system have been introduced and may be considered under three main headings.

Fully pressurised
The tanks are in the form of pressure vessels, usually cylindrical (Fig. 9.12) designed for a maximum pressure of about 18 bar. No reliquefaction plant is required and no insulation is fitted. Relief valves protect the tank against excess

pressure. A compressor is fitted to pressurise the cargo. This tank system has not proved popular due to the considerable loss in hold capacity, the weight of the system and the subsequent cost. Fully pressurised tanks have usually been fitted to small ships having a capacity of less than 2000 m³. The tanks must be strongly-built and are termed *self-supporting*.

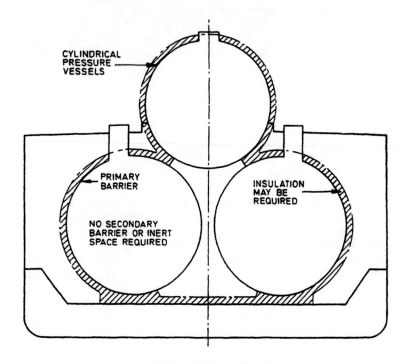

CYLINDRICAL TANK SYSTEM

Fig. 9.12

Semi-pressurised/partly refrigerated

In order to reduce the cost of the tank system, the tank boundaries may be insulated and a reliquefaction plant fitted. The maximum pressure is about 8 bar and the minimum temperature about −5°C. This system has been used on vessels up to about 5000 m³ capacity. The tank system is similar in form to the fully pressurised tanks and has the same loss in hold capacity.

Semi-pressurised/fully refrigerated

This tank system is designed to accept pressures of about 8 bar, but is built of material which will accept temperatures down to about −45°C. The structure must be well-insulated and a reliquefaction plant is necessary. The tanks may be cylindrical or spherical (Fig. 9.13) and are self-supporting. Ships of this type may carry cargoes under a range of conditions, from high pressure at ambient temperature to low temperature at atmospheric pressure.

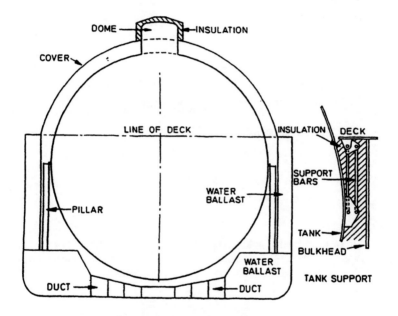

SPHERICAL TANK SYSTEM
Fig. 9.13

Fully refrigerated

Cargo is carried at atmospheric pressure and at a temperature at or below the boiling point. The system is particularly suitable for the carriage of LNG but is also used extensively for LPG and ammonia.

Vessels designed for LNG do not usually carry reliquefaction plant but LPG and ammonia vessels may gain from its use.

The tank structure may be of prismatic form (Fig. 9.14) or of membrane construction.

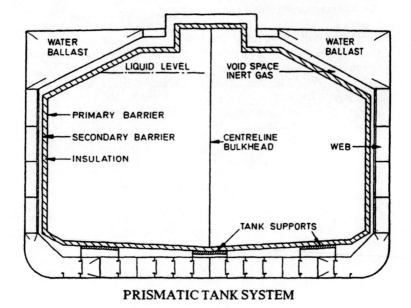

PRISMATIC TANK SYSTEM

Fig. 9.14

Prismatic tanks are self-supporting, being tied to the main hull structure by a system of chocks and keys (Fig. 9.15). They make excellent use of the available space.

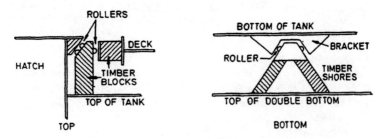

TANK CHOCKS

Fig. 9.15

Membrane tanks are of rectangular form and rely on the main hull structure for their strength. A very thin lining (0.5 mm to 1.2 mm) contains the liquid. This lining must be constructed of low expansion material or must be of corrugated form to allow for changes in temperature. The lining is supported by insulation which must therefore be load-bearing (Fig. 9.16).

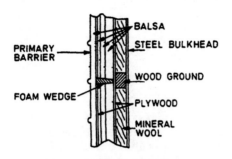

MEMBRANE SYSTEM

Fig. 9.16

For both membrane and prismatic tanks having a minimum temperature less than − 50°C either nickel steel or aluminium must be used. In both cases secondary barriers are required if the minimum temperature is less than − 10°C. Thus, in the event of leakage from the primary container, the liquid or vapour is contained for a period of up to 15 days. If the minimum temperature is higher than − 50°C, the ship's hull may be used as a secondary barrier if constructed of Arctic D steel or equivalent. Independent secondary barriers may be of nickel steel, aluminium or plywood as long as they can perform their function.

Several types of insulation are acceptable, such as balsa wood, polyurethane, pearlite, glass wool and foam glass (Fig. 9.17). Indeed, the primary barrier itself may be constructed of polyurethane which will both contain and insulate the cargo. Usually, however, the primary barrier is of low-temperature steel or of aluminium, neither of which become brittle at low temperatures.

Care must be taken throughout the design to prevent the low-temperature liquid or vapour coming into contact with steel structure which may become brittle and hence fracture.

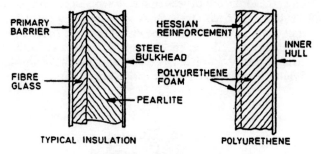

TANK INSULATION

Fig. 9.17

Safety and environmental control

Due to variations in external conditions and the movement of the ship the liquid cargo will boil and release gas. This gas must be removed from the tank to avoid increased pressure. If the gas is lighter than air it may be vented to the atmosphere, vertically from the ship and away from the accommodation. Many authorities are concerned about the increased pollution and alternatives are encouraged (see 'Boil-off').

The cargo piping system is designed for easy gas freeing and purging, with gas sampling points for each tank. The spaces between the barriers and between the secondary barrier and the ship side are either constantly inerted or sufficient inert gas is made available if required.

Each tank is fitted with means of indicating the level of the liquid, the pressure within the tank and the temperature of the cargo. A high liquid level alarm is fitted, giving both audible and visible warning and automatically cuts off the flow.

High and low pressure alarms are fitted within the tanks and in the inter-barrier space if that space is not open to the atmosphere. A temperature measuring device is fitted near the top and at the bottom of each tank with an indicator showing the lowest temperature for which the tank is designed. Temperature readings are recorded at regular intervals, whilst an alarm will be given if the minimum temperature is approached.

Gas detection equipment is fitted in inter-barrier spaces, void spaces, cargo pump rooms and control rooms. The type of equipment depends upon the type of cargo and the space in question. Measurements of flammable vapour, toxic vapour, vapour and oxygen content are taken, audible and visible alarms being actuated if dangerous levels are recorded. Measurements for toxic gas are recorded every four hours except when personnel are in the compartment, when 30 min samples are taken and analysed.

In the event of a fire, it is essential that the fire pumps are capable of supplying two jets or sprays which can reach all parts of the deck over the cargo tanks. The main fire pumps or a special spray pump may be used for protecting the cargo area. The sprays may also be used to reduce the deck temperature during the voyage and hence reduce the heat gain by the cargo.

If the cargo is flammable, then a fixed dry chemical fire extinguishing system is fitted.

Boil off

In the early stage of LNG design, excess methane gas was vented astern of the ship and burned. Reliquefaction is not economical, but the exhaust gas may be used as fuel for the main engine. In motor vessels normal injection equipment is used, together with a hydraulically-operated gas injection valve in the cylinder head, blowing combustion gas at about 3 bar against the incoming scavenge air. The gas line from the tanks is fitted with a relief valve. Under normal running conditions the gas is used as the fuel. Should the gas pressure fall, however, liquid fuel is injected, a 10% drop in pressure resulting in automatic oil supply, whilst if the gas pressure falls by 15% the gas flow is stopped. Oil fuel must be used when starting or manoeuvring.

In steam ships the firing equipment is capable of burning oil fuel and methane gas simultaneously. The oil flame must be present at all times and an alarm system is fitted to give warning of pump failure causing loss of oil. The nozzles are purged with inert gas or steam before and after use. As with motor ships oil fuel is used when starting or manoeuvring.

The excess gas in LPG ships is either vented to the atmosphere or reliquefied by passing it through a cooling system and returning it to the tanks in liquid form.

Operating procedures

Drying

Water in any part of the cargo handling system will impair its operation, by freezing, reducing the purity of the cargo or in some cases changing its nature. The system must be cleared of water or water vapour by purging with a dry gas or by the use of a drying agent.

Inerting

If the oxygen content in a tank is too high a flammable mixture may be produced or the oxygen may be absorbed by the cargo producing a chemical change. It is essential, therefore, to reduce the oxygen content by the introduction of inert gas to a maximum content of 6% for hydrocarbons, 12% for ammonia and 0.5% for ethylene. The most suitable inert gas is nitrogen but this is expensive. Inert gas generated by ship-board plant usually consists of about 84% nitrogen and 12% to 15% carbon dioxide with an oxygen content of about 0.2%.

The inert gas, in turn, must be purged from the system by the cargo gas vapour. An inert gas barrier must also be used when discharging cargo and allowing air into the tanks.

Pre-cooling

Classification Societies require that the maximum temperature difference between the cargo and the tank should not exceed 28°C. Before loading, the tank may be at ambient temperature. It is then cooled by spraying liquefied gas into the tank. The gas then vapourises and cools down the tank. The vapour produced in this way is either vented or reliquefied. The cooling rate must be controlled to prevent undue thermal stresses and excess vapour and a rate of between 3°C and 6°C per hour is usual.

Loading and discharging

The cargo pipeline must first be cooled before loading commences. The rate of loading depends largely on the rate at which the cargo vapour can be vented or reliquefied. Thus a ship designed for a particular run should have a reliquefaction plant compatible with the loading facilities.

When the cargo is discharged it must be in a condition suitable for the shore-based tanks. Thus, if the shore tanks are at atmospheric pressure, the cargo in the ship's tanks should be brought to about the same pressure before discharging commences.

Sufficient nett positive suction head must be provided for the pumps to work the cargo without cavitation. In pressure vessels the ship's compressor may pressurise the vapour from an adjoining tank to maintain a positive thrust at the impeller. In refrigerated ships the impeller is fitted at the bottom of the tank, the liquid head producing the pressure required. In the event of pump failure, the cargo may be removed by the injection of inert gas. Booster pumps are usually fitted to overcome any individual pump problem and ensure a continuous rate of discharge.

CONTAINER SHIPS

Container ships are designed to carry large numbers of standard containers at high speeds between particular terminal ports and require a fast turn round at those ports.

The containers are of international standard, 20 ft, 30 ft or 40 ft in length, 8 ft wide and 8 ft high, the 20 ft and 40 ft lengths being most popular. They are strong enough to be stacked six high. Two basic types of refrigerated container are available, one which carries its own refrigeration plant, either fixed or clipped on, and one which relies on air from brine coolers in the ship which is ducted to the container.

The containers are loaded into the ship vertically, fitting into cell guides which are splayed out at the top to provide lead-in. Pads are fitted to the tank top at the bottom of the guides in line with the corner fittings. The available hold space is dictated by the size of the hatches. It is essential, therefore, to have long, wide hatches to take a maximum number of containers. The spaces at the sides of the hatch are used for access and water ballast. The hatch coamings and covers are designed to carry tiers of containers as deck cargo. Since the vessels usually work

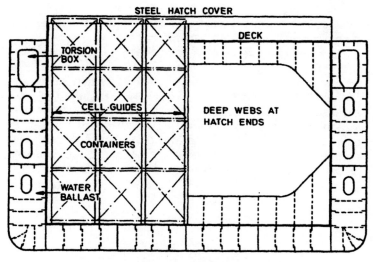

CONTAINER SHIP

Fig. 9.18

between well-equipped ports, they do not usually carry their own cargo handling equipment.

Because of the wide hatches the deck plating must be thick, and higher tensile steel is often used. The deck, side shell and longitudinal bulkheads are longitudinally framed in addition to the double bottom. The hatch coamings may be continuous and therefore improve the longitudinal strength. Problems may arise in these vessels due to the lack of torsional strength caused by the large hatches. This problem is overcome to some extent by fitting *torsion boxes* on each side of the ship. These boxes are formed by the upper deck, top part of the longitudinal bulkhead, sheerstrake and upper platform, all of which are of thick material. The boxes are supported inside by transverses and wash bulkheads in addition to the longitudinal framing. These boxes are only effective if they are efficiently tied at their ends. At the after end they extend into the engine room and are tied to deep transverse webs. Similarly at the fore end, they are carried as far forward as the form of the ship will allow and are welded to transverse webs. The longitudinal bulkheads below the box may have to be stepped inboard to suit the shape of the ship, the main longitudinal bulkhead being scarphed into the stepped section.

At the ends of the hatches deep box webs are fitted to increase the transverse and torsional strength of the ship. These webs are fitted at tank top and deck levels. Care is taken in the structural design at the hatch corners to avoid excessive stresses.

The double bottom structure beneath the cell guides is subject to impact loading as the containers are put on board. Side girders are usually fitted under the container seats with additional transverse local stiffening to distribute the load. Unlike normal cargo ships in which the cargo is distributed over the tank top, the inner bottom of a container ship is subject to point loading. The double bottom must be deep enough to support the upthrust from the water when the ship is deeply loaded, without distortion between the container corners.

CHAPTER 10

FREEBOARD, TONNAGE, LIFE SAVING APPLIANCES, FIRE PROTECTION AND CLASSIFICATION

FREEBOARD

Freeboard is the distance from the waterline to the top of the deck plating at the side of the freeboard deck amidships. The *freeboard deck* is the uppermost continuous deck having means of closing all openings in its weather portion.

The *minimum freeboard* is based on providing a volume of reserve buoyancy which may be regarded as safe, depending upon the conditions of service of the ship. In deep sea ships, sufficient reserve buoyancy must be provided to enable the vessel to rise when shipping heavy seas.

The Load Line Rules give a *tabular freeboard* which depends upon the type of ship, the length of the ship and is based on a standard vessel having a block coefficient of 0.68, length ÷ depth of 15 and a standard sheer curve. Corrections are then made to this value for variations from the standard, together with deductions for the reserve buoyancy afforded by weathertight superstructures on the freeboard deck. One further point to consider is the likelihood of water coming onto the fore deck. This is largely a function of the distance of the fore end of the deck from the waterline. For this reason a *minimum bow height* is stipulated. This value depends upon the length of the ship and the block coefficient and may be measured to the forecastle deck if the forecastle is 7% or more of the ship's length. Should the bow height be less than the minimum then either the freeboard is increased or the deck raised by increasing the sheer or fitting a forecastle.

Two basic types of ship are considered:

Type A ships are designed to carry only liquid cargoes and hence have a high integrity of exposed deck, together with excellent subdivision of the cargo space. Because the hatches are oiltight, and heavy seas are unlikely to cause flooding of cargo space or accommodation, these vessels are allowed to load to a comparatively deep draught.

While these ships have a high standard of watertight deck, they have a comparatively small volume of reserve buoyancy and may therefore be unsafe if damaged. It is necessary, therefore, in all such vessels over 150 m in length, to investigate the effect of damaging the underwater part of the cargo space and, in longer ships, the engine room. Under such conditions the vessel must remain afloat without excessive heel and have positive stability.

Type B ships cover the remaining types of vessel and are assumed to be fitted with steel hatch covers. In older ships having wood covers the freeboard is increased.

Should the hatch covers in Type B ships be gasketted with efficient securing arrangements, then their improved integrity is rewarded by a reduction in freeboard of 60% of the difference between the Type A and Type B tabular freeboards. If, in addition, the vessel satisfies the remaining conditions for a Type A ship (*e.g.* flooding of cargo spaces and engine room), 100% of the difference is allowed and the vessel may be regarded as a Type A ship.

The tabular freeboards for Types A and B ships are given in the Rules for lengths of ship varying between 24 m and 365 m. Typical values are as follows:

Length of ship m	Type A	Type B	Difference
	Freeboard in mm		
24	200	200	—
100	1135	1271	136
200	2612	3264	652
300	3262	4630	1368
365	3433	5303	1870

The freeboard calculated from the tabular freeboard and corrected is termed the Summer Freeboard and corresponds with a Summer Load Line S.

In the tropics the weather is usually kind and a deduction of $\frac{1}{48}$ of the Summer draught may be made to give the Tropical Load Line T.

Similarly there is a Winter Load Line W which is a penalty of $\frac{1}{48}$ of the Summer draught. In ships 100 m and less in length there is a further penalty of 50 mm if the vessel enters the North Atlantic in the Winter (WNA).

The above freeboards are based on the assumption that the

ship floats in sea water of 1025 kg/m³. If the vessel floats in fresh water with the same displacement, then it will lie at a deeper draught. The *Fresh Water Allowance is* $\frac{\Delta}{4T}$ mm where Δ is the displacement in sea water at the Summer draught and T is the tonne per cm at the same draught. F represents the fresh water line in the Summer zone and TF the equivalent mark in the Tropical zone.

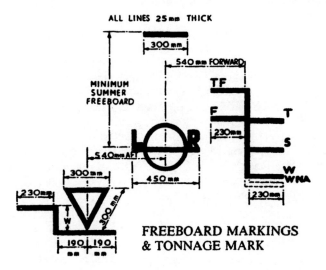

FREEBOARD MARKINGS
& TONNAGE MARK

Fig. 10.1

The freeboard markings (Fig. 10.1) are cut into the shell plating with the centre of the circle at midships. The letters LR on the circle indicate that the load line has been assigned by Lloyd's Register of Shipping. If the vessel has a radiused gunwale, the deck line is cut at a convenient distance below the correct position and this distance is then deducted from the freeboard stated on the Load Line Certificate.

Special provision is made in the Rules for vessels carrying timber as a deck cargo. The timber increases the reserve buoyancy and hence the vessels are allowed to float at deeper draughts. An additional set of freeboard markings is cut in aft of midships with the normal letters prefixed by L (lumber).

Conditions of Assignment

The Load Line Rules are based on the very reasonable assumption that the ship is built to and maintained at a high level of structural strength and will sail in a safe and seaworthy condition.

Until recently the Rules laid down the standard of longitudinal and transverse strength. The Classification Societies usually found it necessary to increase these standards although in some designs considered the Rules excessive. It is now felt that the structural strength of the ship is more properly the function of the Classification Societies who may well be the Assigning Authority.

Standards of stability are given in the Rules for both small and large angles of heel. Details of the information required to be carried on a ship are stated, together with typical calculations. All the information is based on an inclining experiment carried out on the completed ship in the presence of a D.Tp. surveyor.

It is essential that all openings in the weather deck are weathertight. Hatch coamings, hatch covers, ventilator coamings, air pipes and doors must be strong enough to resist the pounding from the sea and standards of strength are laid down. The Rules also specify the height of coamings, air pipes and door sills above the weather deck, those at the fore end being higher than the remainder.

It is important to remove the water from the deck quickly when a heavy sea is shipped. With completely open decks, the reserve buoyancy is sufficient to lift the ship and remove the water easily. When bulwarks are fitted, however, they tend to hold back the water and this may prove dangerous. For this reason openings known as freeing ports are cut in the bulwarks, the area of the freeing ports depending upon the length of the bulwark. If the freeing ports are wide, grids must be fitted to prevent crew being washed overboard. In addition, scuppers are fitted to remove the surplus water from the deck. The scuppers on the weather deck are led overboard whilst those on intermediate decks may be led to the bilges or, if automatic non-return valves are fitted, may be led overboard.

Type A ships, with their smaller freeboard, are more likely to have water on the deck and it is a condition of assignment that open rails be fitted instead of bulwarks. If the vessel has midship accommodation, a longitudinal gangway must be fitted to allow passage between the after end and midships without setting foot on the weather deck. In larger ships it is necessary to fit shelters along the gangway. Alternatively access may be provided by an underdeck passage, but while convenient for bulk carriers could prove dangerous in oil tankers.

Surveys

To ensure that the vessel is maintained at the same standard of safety, annual surveys are made by the Assigning Authority. An inspection is made of all those items which affect the freeboard of the ship and are included in the Conditions of Assignment. The accuracy of the freeboard marks is checked and a note made of any alterations to the ship which could affect the assigned freeboard.

TONNAGE

1967 Rules

In 1967 the Tonnage Rules were completely revised in an attempt to improve the safety of dry cargo ships.

A *register ton* represents 100 cubic feet of volume*. The *tonnage deck* is the second deck except in single deck ships. The *tonnage length* is measured at the level of the tonnage deck where an imaginary line is drawn inside the hold frames or sparring, the tonnage length being measured on the centreline of the ship to this line. *Tonnage depths* are measured from the top of the tank top or ceiling to the underside of the tonnage deck at the centreline, less one third of the camber. There is, however, a limitation on the height of the double bottom considered. *Tonnage breadths* are measured to the inside of the hold frames or sparring.

The tonnage length is divided into a number of parts. At each cross-section the tonnage depth is similarly divided and tonnage breadths measured. The breadths are put through Simpson's Rule to give cross-sectional areas. The cross-sectional areas are, in turn, put through Simpson's Rule to give a volume. This volume, divided by 100, is the *Underdeck Tonnage.*

The *Gross Tonnage* is found by adding to the Underdeck Tonnage, the tonnage of all enclosed spaces between the upper deck and the second deck, the tonnage of all enclosed spaces above the upper deck together with any portion of hatchways exceeding $\frac{1}{2}\%$ of the gross tonnage.

The *Net Tonnage* or *Register Tonnage* is obtained by deducting from the Gross Tonnage, the tonnage of spaces which are required for the safe working of the ship:

(a) master's accommodation
(b) crew accommodation and an allowance for provision stores

*This has not changed with the introduction of SI units

(c) wheelhouse, chartroom, radio room and navigation aids room
(d) chain locker, steering gear space, anchor gear and capstan space
(e) space for safety equipment and batteries
(f) workshops and storerooms for pumpmen, electricians, carpenter, boatswains and the lamp room
(g) donkey engine and donkey boiler space if outside the engine room
(h) pump room if outside the engine room
(i) in sailing ships, the storage space required for the sails, with an upper limit of $2\frac{1}{2}\%$ of the gross tonnage
(j) water ballast spaces if used only for that purpose. The total deduction for water ballast, including double bottom spaces, may not exceed 19% of the gross tonnage
(k) *Propelling Power Allowance.* This forms the largest deduction and is calculated as follows:

If the Machinery Space Tonnage is between 13% and 20% of the Gross Tonnage, the Propelling Power Allowance is 32% of the Gross Tonnage.

If the Machinery Space Tonnage is less than 13% of the Gross Tonnage, the Propelling Power Allowance is a proportion of 32% of the Gross Tonnage. Thus an actual tonnage of 12% would give a Propelling Power Allowance of $\frac{12}{13} \times 32\%$ of the Gross Tonnage.

If the Machinery Space Tonnage is more than 20% of the Gross Tonnage, the Propelling Power Allowance is $1\frac{3}{4}$ times the Machinery Space Tonnage, with an upper limit of 55% except for tugs.

Modified tonnage
Many ships are designed to run in service at a load draught which is much less than that allowed by the Load Line Rules. If the freeboard of a vessel is greater than that which would be assigned taking the second deck as the freeboard deck then reduced Gross and Net Tonnages may be allowed. In this case the tonnage of the space between the upper deck and the second deck is not added to the Underdeck Tonnage and is therefore not included in the Gross Tonnage or Net Tonnage, both of which are consequently considerably reduced. As an indication that this *modified tonnage* has been allocated to the ship, a *tonnage mark* must be cut in each side of the ship in line with the deepest loadline and 540 mm aft of the centre of the disc.

If any cargo is carried in the 'tween decks it is classed as deck cargo and added to the tonnage.

Alternative tonnage

The owner may, if he wishes, have assigned to any ship reduced Gross and Net Tonnages calculated by the method given above. This is an *alternative* to the normal tonnages and is penalised by a reduction in the maximum draught. A *tonnage mark* must be cut in each side of the ship at a distance below the second deck depending upon the ratio of the tonnage length to the depth of the second deck. If the ship floats at a draught at or below the apex of the triangle, then the reduced tonnages may be used. If, however, the tonnage mark is submerged, then the normal tonnages must be used.

The principle behind the modified and alternative tonnages is that reduced tonnages were previously assigned if a tonnage hatch were fitted. This hatch seriously impaired the safety of the ship. Thus by omitting the hatch the ship is more seaworthy and no tonnage penalty is incurred. The tonnage mark suitable for alternative tonnage is shown in Fig. 10.1. The distance W is $\frac{1}{48} \times$ the moulded draught to the tonnage mark.

1982 Rules

The 1967 and earlier Tonnage Rules influenced the design of ships and introduced features which were not necessarily consistent with the safety and efficiency of the ship. In 1969 an International Convention on Tonnage Measurement of ships was held and new Tonnage Rules were produced. These Rules came into force in 1982 for new ships, although the 1967 Rules could still be applied to existing ships until 1994.

The principle behind the new Rules was to produce similar values to the previous rules for gross and net tonnage using a simplified method which reflected more closely the actual size of the ship and its earning capacity without influencing the design and safety of the ship.

The gross tonnage is calculated from the formula

Gross tonnage (GT)

$$= K_1 V$$

where　　　　V = total volume of all enclosed spaces in the ship in m^3

$K_1 = 0.2 + 0.02 \log_{10} V$

Thus the gross tonnage depends upon the total volume of the ship and therefore represents the size of the ship.

Enclosed spaces represent all those spaces which are bounded by the ship's hull, by fixed or portable partitions or bulkheads and by decks or coverings other than awnings.

Spaces excluded from measurement are those at the sides and ends of erections which cannot be closed to the weather and are not fitted with shelves or other cargo fitments.

The net tonnage is claculated from the formula

$$\text{Net tonnage (NT)} = K_2 V_C \left(\frac{4d}{3D}\right)^2 + K_3 \left(N_1 + \frac{N_2}{10}\right)$$

where
V_C = total volume of cargo spaces in m³
$K_2 = 0.2 + 0.02 \log_{10} V_C$
$K_3 = 1.25 \left(\dfrac{GT + 10\ 000}{10\ 000}\right)$
D = moulded depth amidships in m
d = moulded draught amidships in m
N_1 = number of passengers in cabins with not more than 8 berths
N_2 = number of other passengers

When a ship is designed to carry less than 13 passengers, the second term in the equation is ignored and the net tonnage is then based directly on the cargo capacity.

There are three further conditions:

The term $\left(\dfrac{4d}{3D}\right)$ is not to be taken as greater than unity.

The term $K_2 V_C \left(\dfrac{4d}{3D}\right)^2$ is not to be taken as less than 0.25 GT.

The net tonnage is not to be taken as less than 0.30 GT.

Hence ships which carry no passengers and little or no cargo will have a net tonnage of 30% of the gross tonnage.

All cargo spaces are certified by permanent markings CC (cargo compartment).

The result of these rules will be that Shelter Deck Vessels carrying dual tonnage and the Tonnage Mark will disappear from the scene.

A unified system of measurement will be used by all nations with no variations in interpretation.

Net tonnage is used to determine canal dues, light dues, some pilotage dues and some harbour dues.

Gross tonnage is used to determine manning levels, safety requirements such as fire appliances, some pilotage and harbour dues, towing charges and graving dock costs.

LIFE-SAVING APPLIANCES

The life-saving equipment carried on board a ship depends upon the number of persons carried and the normal service of the ship. A Transatlantic passenger liner would carry considerably more equipment than a coastal cargo vessel. The following notes are based on the requirements for a deep-sea cargo ship.

There must be sufficient lifeboat accommodation on *each* side of the ship for the whole of the ship's complement. The lifeboats must be at least 7.3 m long and may be constructed of wood, steel, aluminium or fibreglass. They carry rations for several days, together with survival and signalling equipment such as fishing lines, first-aid equipment, compass, lights, distress rockets and smoke flares. One lifeboat on each side must be motor-driven.

The lifeboats are suspended from davits which allow the boats to be lowered to the water when the ship is heeled to 15 degrees. Most modern ships are fitted with gravity davits, which, when released, allow the cradle carrying the boat to run outboard until the boat is hanging clear of the ship's side (Fig. 10.2).

The boat is raised and lowered by means of an electrically-driven winch. The winch is manually controlled by a weighted lever (Fig. 10.3), known as a *dead man's handle* which releases the main brake. Should the operator lose control of the brake the lever causes the winch to stop. The speed of descent is also controlled by a centrifugal brake which limits the speed to a maximum of 36 m/min. Both the centrifugal brake and the main brake drum remain stationary during the hoisting operation. If the power fails while raising the boat, the main brake will hold the boat.

The majority of ships carry liferafts having sufficient capacity for half of the ship's crew. The liferafts are inflatable and carry survival equipment similar to the lifeboats. They have been found extremely efficient in practice, giving adequate protection from exposure.

Each member of the crew is supplied with a lifejacket which is capable of supporting an unconscious person safely.

Lifebuoys are provided in case a man falls overboard. Some are fitted with self-igniting lights for use at night and others fitted with smoke signals for pin-pointing the position during the day.

All ships carry line throwing apparatus which consists of a light line to which a rocket is attached. The rocket is fired from a

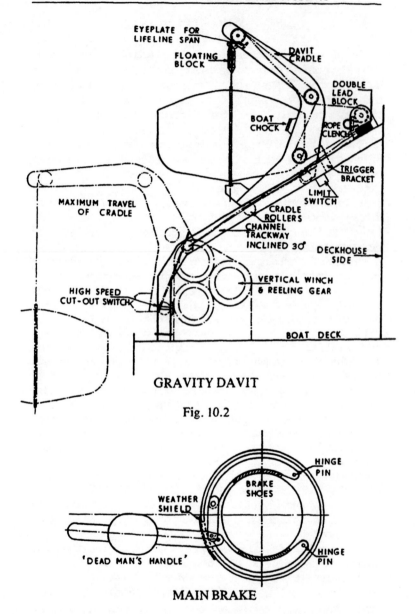

GRAVITY DAVIT

Fig. 10.2

MAIN BRAKE

Fig. 10.3

pistol and must be capable of carrying 230 m. This enables
contact to be made between the ship and the shore or another

ship. A hawser or whip is attached to the end of the line and pulled back onto the ship either directly or through a block, allowing persons to be transferred or vessels towed.

FIRE PROTECTION

Definitions

A *non-combustible material* is one which neither burns nor gives off flammable vapours in sufficient quantity for self-ignition when heated to 750°C.

A *standard fire test* is carried out on a stiffened panel of material 4.65 m² in area, 2.44 m high with one joint. One side of the panel is exposed in a test furnace to a series of temperatures, 538°C, 704°C, 843°C and 927°C at the end of 5, 10, 30 and 60 minute periods respectively.

An *A-class division* must be made of steel or equivalent material capable of preventing the passage of smoke and flame to the end of the 60 minute standard fire test. The average temperature on the unexposed surface of the panel must not rise more than 139°C after a given time and it may be necessary to insulate the material. The time intervals are 0, 15, 30 and 60 minutes and the divisions classed to indicate this interval *i.e.* A-0; A-60.

A *B-class division* need not be made of steel but must be of non-combustible material and must prevent the passage of smoke and flame to the end of the first 30 minutes of the standard fire test. The average temperature on the unexposed surface must not rise more than 139°C after 0 or 15 minutes.

A *C-class division* must be of non-combustible material.

Control stations refer to spaces containing main navigation or radio equipment, central fire-recording system or the emergency generator.

Fire potential is the likelihood of fire starting or spreading in a compartment. If the fire potential is high, then a high standard of insulation is required. Thus bulkheads separating accommodation from machinery spaces would be required to be A-60 whilst those dividing accommodation from sanitary spaces could be B-0 or even C.

There are three basic principles of fire protection:
   (i) the separation of accommodation spaces from the rest of the ship by thermal and structural boundaries.
   (ii) the containment, extinction or detection of fire in the space of origin, together with a fire alarm system.
   (iii) the protection of a means of escape.

Passenger ships

Passenger ships are divided into main vertical zones by A-class divisions not more than 40 m apart. These divisions are carried through the main hull, superstructure and deckhouses. If it is necessary to step the bulkhead, then the deck within the step must also be A-class.

The remainder of the bulkheads and decks within the main vertical zones are A, B or C class depending upon the fire potential and relative importance of the adjacent compartments. Thus bulkheads between control stations or machinery spaces and accommodation will be A-60 whilst those between accommodation and sanitary spaces will be B-0.

The containment of fire vertically is extremely important and the standard of protection afforded by the decks is similar to that of the bulkheads.

If a sprinkler system is fitted the standard of the division may be reduced, typically from A60 to A-15.

All compartments in the accommodation, service spaces and control stations are fitted with automatic fire alarm and detection systems.

All stairways are of steel or equivalent and are within enclosures formed by A-class divisions. Lift trunks are designed to prevent the passage of smoke and flame between decks and to reduce draughts. Ventilator trunks and ducts passing through main vertical zone bulkheads are fitted with dampers capable of being operated from both sides of the bulkheads.

Fire resisting doors may be fitted in the A-class bulkheads forming the main vertical zone and those enclosing the stairways. They are usually held in the open position but close automatically when released from a control station or at the door position even if the ship is heeled ± 3.5°.

Vehicle spaces in ships having drive on/drive off facilities present particular problems because of their high fire potential and the difficulty of fitting A-class divisions. A high standard of fire extinguishing is provided by means of a *drencher* system. This comprises a series of full-bore nozzles giving an even distribution of water of between 3.5 and 5.0 l/m²min. over the full area of the vehicle deck. Separate pumps are provided for the system.

Dry cargo ships

In ships over 4000 tons gross all the corridor bulkheads in the accommodation are of steel or B-class. The deck coverings inside accommodation which lies above machinery or cargo spaces must not readily ignite. Interior stairways and crew lift trunks are of steel as are bulkheads of the emergency generator room and bulkheads separating the galley, paint store, lamp room or bosun's store from accommodation.

Oil Tankers

In tankers of over 500 tons gross the machinery space must lie aft of the cargo space and must be separated from it by a cofferdam or pumproom. Similarly all accommodation must lie aft of the cofferdam. The parts of the exterior of the superstructure facing the cargo tanks and for 3 m aft must be A-60 standard. Any bulkhead or deck separating the accommodation from a pump room or machinery space must also be of A-60 standard.

Within the accommodation the partition bulkheads must be of at least C standard. Interior stairways and lift trunks are of steel, within an enclosure of A-0 material.

To keep deck spills away from the accommodation and service area a permanent continuous coaming 150 mm high is welded to the deck forward of the superstructure.

It is important to prevent gas entering the accommodation and engine room. In the first tier of superstructure above the upper deck no doors to accommodation or machinery spaces are allowed in the fore end and for 5 m aft. Windows are not accepted but non-opening ports may be fitted but are required to have internal steel covers. Above the first tier non-opening windows may be fitted in the house front with internal steel covers.

CLASSIFICATION OF SHIPS

A classification society is an organisation whose function is to ensure that a ship is soundly constructed and that the standard of construction is maintained. The ship is classified according to the standard of construction and equipment. The cost of insurance of both ship and cargo depends to a great extent upon this classification and it is therefore to the advantage of the shipowner to have a high class ship. It should be noted, however, that the classification societies are independent of the insurance companies.

There are a number of large societies, each being responsible for the classification of the majority of ships built in at least one country, although in most cases it is left to the shipowner to choose the society. Some of the major organisations are as follows:

Lloyd's Register of Shipping	United Kingdom
American Bureau of Shipping	U.S.A.
Bureau Veritas	France
Det Norske Veritas	Norway
Registro Italiano	Italy
Teikoku Kaiji Kyokai	Japan
Germanischer Lloyds	Germany

Each of these societies has its own rules which may be used to determine the scantlings of the structural members. The following notes are based on Lloyd's Rules.

Steel ships which are built in accordance with the Society's Rules, or are regarded by Lloyds as equivalent in strength, are assigned a class in the Register Book. This class applies as long as the ships are found under survey to be in a fit and efficient condition.

Class 100A is assigned to ships which are built in accordance with the rules or are of equivalent strength.

The figure 1 is added (*i.e.*, 100A 1) when the equipment, consisting of anchors, cables, mooring ropes and towropes, is in good and efficient condition.

The distinguishing mark ✠ is given when a ship is fully built under Special Survey, *i.e.*, when a surveyor is in attendance and examines the ship during all stages of construction. Thus a ship classed as ✠ 100A 1 is built to the highest standard assigned by Lloyds.

Additional notations are added to suit particular types of ship such as 100A 1 oil tanker or 100A 1 ore carrier.

When the machinery is constructed and installed in accordance with Lloyd's Rules a notation LMC is assigned, indicating that the ship has Lloyd's Machinery Certificate.

In order to claim the 100A 1 class, the materials used in the construction of the ship must be of good quality and free from defects. To ensure that this quality is obtained, samples of material are tested at regular intervals by Lloyd's Surveyors.

To ensure that the ship remains worthy of its classification, annual and special surveys are carried out by the surveyors. The special surveys are carried out at intervals of 4 to 5 years.

In an annual survey the ship is examined externally and, if considered necessary, internally. All parts liable to corrosion and chafing are examined, together with the hatchways, closing devices and ventilators to ensure that the standards required for the Load Line Regulations are maintained. The steering gear, windlass, anchors and cables are inspected.

A more thorough examination is required at the special surveys. The shell plating, sternframe and rudder are inspected, the rudder being lifted if considered necessary. The holds, peaks, deep tanks and double bottom tanks are cleared, examined and the tanks tested. The bilges, limbers and tank top are inspected, part of the tank top ceiling being removed to examine the plating. In way of any corroded parts, the thickness of the plating must be determined either by drilling or by ultrasonic testing.

The scantlings of the structure are based on theory, but because a ship is a very complex structure, a 'factor of experience' is introduced. Lloyds receive reports of all faults and failures in ships which carry their class, and on the basis of these reports, consistent faults in any particular type of ship may be studied in detail and amendments made to the rules. Structural damage in some welded ships has led to the introduction of longitudinal framing in the double bottom. Reports of brittle fracture have resulted in crack arrestors being fitted in the shell and deck of welded ships. It is important to note that Lloyds have the power to require owners to alter the structure of an existing ship if they consider that the structure is weak. An example of this was the fitting of butt straps to some welded tankers to act as crack arrestors, the shell plating being cut to create a discontinuity in the material, and the separate plates joined together by means of the butt strap.

In order that a ship may receive Lloyd's highest classification, scantling plans are drawn. On these plans the thicknesses of all plating, sizes of beams and girders and the method of construction are shown. The scantlings are obtained from Lloyd's Rules and depend upon the length, breadth, draught, depth, and frame spacing of the ship and the span of the members. Variations may occur due to special design characteristics such as the size and position of the machinery space. Shipowners are at liberty to increase any of the scantlings and many do so, particularly where such increases lead to reduced repair costs. An increase in diameter of rudder stock by 10% above rule is a popular owners' extra. Shipbuilders may also submit alternative arrangements to those given in the rules and Lloyds may allow their use if they are equivalent in strength. The scantling plans are submitted to Lloyds for their approval before detail plans are drawn and the material ordered.

One major problem which has arisen is the increase in the size of oil tankers and bulk carriers with the consequent small number of docks capable of handling them. This has led to the introduction of a system of hull survey while the vessel is afloat. The in-water survey (IWS) is used to check those parts of a ship which are usually surveyed in dry dock. It includes visual examination of the hull, rudder, propeller, sea inlets, etc., and the measurement of wear down of rudder bearings and stern bush.

There are several requirements to be met before IWS is allowed. Colour photographs are taken of all parts which are likely to be inspected, before the ship is launched. The rudder and sternframe are designed for easy access to bearings. The ship must be less than 10 years old, have a high resistance paint on the underwater hull and be fitted with an impressed current cathodic protection system.

At the time of the inspection the hull is cleaned by one of the many brush systems available. The water must be clear and the draught less than 10 m. The inspection may be made by an underwater closed-circuit television camera. The camera may be hand-held by a diver or carried by a hydraulically-propelled camera vehicle, remotely controlled from a surface monitoring station.

DISCONTINUITIES

If there is an abrupt change in section in any type of stressed structure, particularly high stresses occur at the discontinuity. Should the structure be subject to fluctuating loads, the likelihood of failure at this point is greatly increased.

A ship is a structure in which discontinuities are impossible to avoid. It is also subject to fluctuations or even reversals of stress when passing through waves. The ship must be designed to reduce such discontinuities to a minimum, while great care must be taken in the design of structural detail in way of any remaining changes in section.

The most highly-stressed part of the ship structure is usually that within 40% to 50% of its length amidships. Within this region every effort must be made to maintain a continuous flow of material.

Difficulties occur at hatch corners. Square corners must be avoided and the corners should be radiused or elliptical. With radiused corners the plating at the corners must be thicker than the remaining deck plating. Elliptical corners are more efficient in reducing the corner stress and no increase in thickness is required.

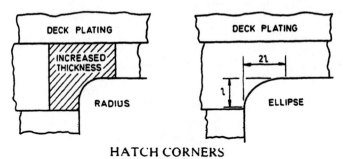

HATCH CORNERS

Fig. 10.4

Similarly openings for doors, windows, access hatches, ladderways, etc., in all parts of the ship must have rounded corners with the free edges dressed smooth.

If a bridge structure is fitted over more than 15% of the ship's length, the bridge side plating must be tapered or curved down to the level of the upper deck. The sheerstrake is increased in thickness by 50% and the upper deck stringer plate by 25% at the ends of the bridge. Four 'tween deck frames are carried through the upper deck into the bridge space at each end of the bridge to ensure that the ends are securely tied to the remaining structure.

In bulk carriers, where a large proportion of the deck area is cut away to form hatches, the hatch coamings should preferably be continuous and tapered down to the deck level at the ends of the ship.

Longitudinal framing must be continued as far as possible into the ends of the ship and scarphed gradually into a transverse framing system. This problem is overcome in oil tankers and bulk carriers by carrying the deck and upper side longitudinals through to the collision bulkhead, with transverse framing in the fore deep tank up to the tank top level and in the fore peak up to the upper deck. Similarly at the after end the side and deck longitudinals are carried aft as far as they will conveniently go. A particular difficulty arises with the longitudinal bulkheads which must be tapered off into the forward deep tank and engine room. In the larger vessels it is often possible to carry the bulkheads through the engine room, the space at the sides being used for auxiliary spaces, stores and workshops.

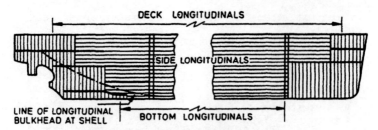

EXTENT OF LONGITUDINAL FRAMING
OIL TANKER

Fig. 10.5

A similar problem occurs in container ships, when it is essential to taper the torsion box and similar longitudinal stiffening gradually into the engine room and the fore end. The fine lines of these ships cause complications which are not found in the fuller vessels. In these faster vessels the importance of tying the main hull structure efficiently into the engine room structure cannot be sufficiently emphasised.

CHAPTER 11

SHIP DYNAMICS

PROPELLERS

This section should be read in conjunction with the chapter on Propellers in Vol. 4 "Naval Architecture for Marine Engineers".

The design of a propulsion system for a ship is required to be efficient for the ship in its intended service, reliable in operation, free from vibration and cavitation, economical in first cost, running costs and maintenance. Some of these factors conflict with others and, as with many facets of engineering, the final system is a compromise. Various options are open to the shipowner including the number of blades, the number of propellers, the type and design of propellers.

Propellers work in an adverse environment created by the varying wake pattern produced by the after end of the ship at the propeller plane. Fig. 11.1 shows a typical wake distribution for a single screw ship in terms of wake fraction w_T (wake speed ÷ ship speed). The high wake fractions indicate that the water is being carried along at almost the same speed as the ship. Thus the propeller is working in almost *dead* water. The lower fractions indicate that the water is almost stationary and therefore has a high speed relative to the propeller. When a propeller blade passes through this region it is more heavily loaded. These variations in loading cause several problems.

In a four-bladed propeller, two of the blades are lightly loaded and two heavily loaded when the blades are in the position shown in Fig. 11.1. When the propeller turns through 90° the situation is reversed. The resulting fluctuations in stress may produce cracks at the root of the blades and vibration of the blades. This fluctuation in loading may be reduced by changing the number of the blades. A three-bladed propeller, for instance will only have one blade fully loaded or lightly loaded at a time,

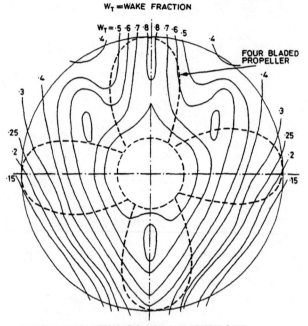

TYPICAL WAKE DISTRIBUTION SINGLE SCREW SHIP

Fig. 11.1
DISTRIBUTION OF PROPELLER WAKE

whilst five, six and seven blades produce more gradual changes in thrust and torque per blade and hence reduce the possibility of vibration due to this cause. It is generally accepted, however, that an increase in blade number results in a reduction in propeller efficiency.

An alternative method of reducing the variation in blade loading is to fit a *skewed* propeller (Fig. 11.2) in which the centreline of each blade is curved to spread the distribution of the blade area over a greater range of wake contours. Such propellers are usually slightly less efficient than the normal propellers but the more gradual change in thrust is thought to reduce stress variations and the possibility of blade vibration.

The thrust of a propeller depends upon the acceleration of a mass of water within its own environment. If a propeller is on the centreline it lies within the wake boundary and accelerates water which is already moving in the same direction as the ship.

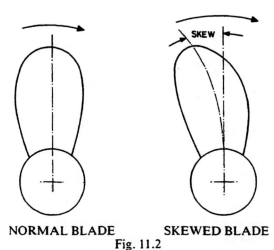

NORMAL BLADE   SKEWED BLADE
Fig. 11.2

A propeller which is off the centreline lies only partly within the wake and does not receive the full benefit from it. For this reason single screw ships are usually more efficient than twin screw ships for similar conditions. The main advantages of twin screw ships are their increased manoeuvrability and the duplication of propulsion systems leading to improved safety. Set against this is the considerable increase in the cost of the construction of the after end, whether the shaft support is by A-frames or by spectacle frames and bossings, compared with the sternframe of a single screw ship.

In a single screw ship the rudder is more effective since it lies in the propeller race and hence the velocity of water at the rudder is increased producing increased rudder force. On the other hand many twin screw ships are fitted with twin rudders in line with the propellers, further increasing their manoeuvrability at the expense of increased cost of steering gear. The variation in wake in a twin screw ship is less than with a single screw ship and the blades are therefore less liable to vibrate due to fluctuations in thrust. The support of the shafting and propeller is less rigid, however, and vibration may occur due to the deflection of the support.

CONTROLLABLE PITCH PROPELLERS

A controllable pitch propeller is one in which the pitch of the blades may be altered by remote control. The blades are

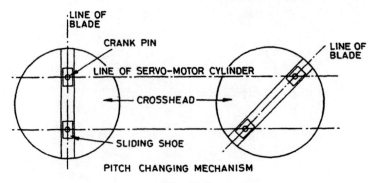

Fig. 11.3

separately mounted onto bearing rings in the propeller hub. A valve rod is fitted within the hollow main shaft and this actuates a servo-motor cylinder. Longitudinal movement of this cylinder transmits a load through a crank pin and sliding shoe to rotate the propeller blade (Fig. 11.3).

The propeller pitch is controlled directly from the bridge and hence closer and quicker control of the ship speed is obtained. This is of particular importance when manoeuvring in confined waters when the ship speed may be changed, and indeed reversed, at constant engine speed. Because full power may be developed astern, the stopping time and distance may be considerable reduced.

The cost of a c.p. propeller is considerably greater than that of a fixed pitch installation. On the other hand a simpler non-reversing main engine may be used or reversing gear box omitted, and since the engine speed may be maintained at all times the c.p. installation lends itself to the fitting of shaft-driven auxiliaries. The efficiency of a c.p. propeller is less than that of a fixed pitch propeller for optimum conditions, due to the larger diameter of boss required, but at different speeds the former has the advantage. The cost of repair and maintenance is high compared with a fixed pitch propeller.

CONTRA-ROTATING PROPELLERS

This system consists of two propellers in line, but turning in opposite directions. The after propeller is driven by a normal solid shaft. The forward propeller is driven by a short hollow shaft which encloses the solid shaft. The forward propeller is usually larger and has a different number of blades from the after propeller to reduce the possibility of vibration due to blade interference.

Research has shown that the system may increase the propulsion efficiency by 10% to 12% by cancelling out the rotational losses imparted to the stream of water passing through the propeller disc.

Contra-rotating propellers are extremely costly and are suitable only for highly loaded propellers and large single screw tankers, particularly when the draught is limited. The increased surface area of the combined system reduces the possibility of cavitation but the longitudinal displacement of the propellers is very critical.

VERTICAL AXIS PROPELLERS

The Voith-Schneider propeller is typical of a vertical axis propeller and consists of a series of vertical blades set into a horizontal rotor which rotate about a vertical axis. The rotor is flush with the bottom of the ship and the blades project down as shown in Fig. 11.4.

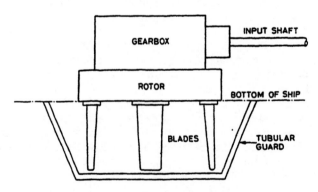

VOITH SCHNEIDER PROPELLER
Fig. 11.4

The blades are linked to a control point P by cranked control rods (Fig. 11.5). When P is in the centre of the disc, the blades rotate without producing a thrust. When P is moved away from the centre in any direction, the blades turn in the rotor out of line with the blade orbit and a thrust is produced. The direction and magnitude of the thrust depends upon the position of P. Since P can move in any direction within its inner circle, the ship

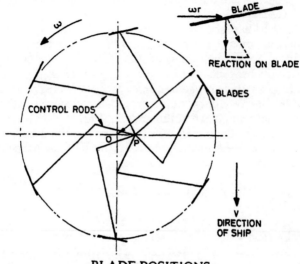

BLADE POSITIONS
Fig. 11.5

may be driven in any direction and at varying speeds. Thus the Voith Schneider may propel and manoeuvre a ship without the use of a rudder.

The efficiency of a Voith Schneider propeller is relatively low but it has the advantage of high manoeuvrability and is useful in harbour craft and ferries. Two or more installations may be fitted and in special vessels (*e.g.*, firefloats), can move the ship sideways or rotate it in its own length.

Replacement of damaged blades is simple although they are fairly susceptible to damage. A tubular guard is usually fitted to protect the blades. The propeller may be driven by a vertical axis motor seated on the top or by a diesel engine with a horizontal shaft converted into vertical drive by a bevel gear unit.

BOW THRUSTERS

Many ships are fitted with bow thrust units to improve their manoeuvrability. They are an obvious feature in ships working within, or constantly in and out of harbour where close control is obtained without the use of tugs. They have also proved to be of considerable benefit to larger vessels such as oil tankers and bulk carriers, where the tug requirement has been reduced.

Several types are available, each having its own advantages and disadvantages.

In all cases the necessity to penetrate the hull forward causes an increase in ship resistance and hence in fuel costs, although the increase is small.

A popular arrangement is to have a cylindrical duct passing through the ship from side to side, in which is fitted an impeller which can produce a thrust to port or to starboard. The complete duct must lie below the waterline at all draughts, the impeller acting best when subject to a reasonable head of water and thus reducing the possibility of cavitation.

The impeller may be of fixed pitch with a variable-speed motor which is reversible or has reverse gearing. Alternatively a controllable pitch impeller may be used, having a constant-speed drive. Power may be provided by an electric motor, a diesel engine or a hydraulic motor.

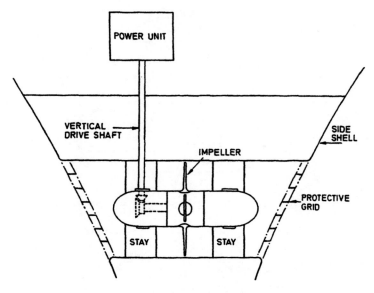

CONTROLLABLE PITCH BOW THRUSTER
Fig. 11.6

Some vessels are fitted with Voith Schneider propellers within the ducts to produce the transverse thrust.

As an alternative the water may be drawn from below the ship and projected port or starboard through a horizontal duct which may lie above or below the waterline. A uni-directional horizontal impeller is fitted in a vertical duct below the waterline. The

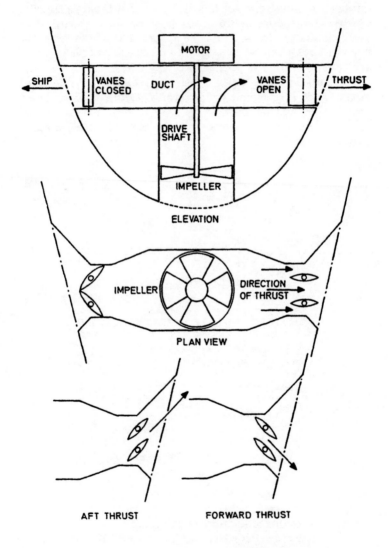

HYDRAULIC THRUST UNIT

Fig. 11.7

lower end of the duct is open to the sea, while the upper end leads into the horizontal duct which has outlets in the side shell port and starboard. Within this duct two hydraulically operated vertical vanes are fitted to each side. Water is drawn from the bottom of the ship into the horizontal duct. By varying the position of the vanes the water jet is deflected either port or starboard, producing a thrust and creating a reaction which pushes the bow in the opposite direction.

This system has an advantage that by turning all vanes 45° either forward or aft an additional thrust forward or aft may be produced. A forward thrust would act as a retarding force whilst an aft thrust would increase the speed of the ship. These actions may be extremely useful in handling a ship in congested harbours.

The efficiency of propeller thruster falls off rapidly as the ship speed increases. The rudder thrust, on the other hand, increases in proportion to the square of the ship speed, being relatively ineffective at low speeds. The water jet unit appears to maintain its efficiency at all speeds, although neither type of thrust unit would normally be used at speed. Fig. 11.8 does indicate the usefulness of thrust units when moving and docking compared with the use of the rudder.

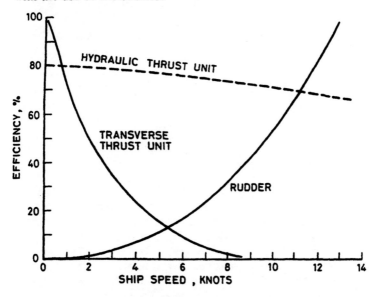

COMPARISON OF EFFICIENCIES

Fig. 11.8

ROLLING AND STABILISATION

When a ship is heeled by an external force, and the force is suddenly removed, the vessel will roll to port and starboard with a rolling period which is almost constant. This is known as the ship's *natural rolling period*. The *amplitude* of the roll will depend upon the applied heeling moment and the stability of the ship. For angles of heel up to about 15° the rolling period does not vary with the angle of roll. The angle reduces slightly at the end of each swing and will eventually dampen out completely. This dampening is caused by the frictional resistance between the hull and the water, which causes a mass of *entrained water* to move with the ship.

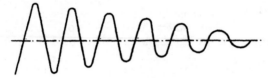

TYPICAL DAMPING CURVE

Fig. 11.9

The natural rolling period of a ship may be estimated by the formula:

$$\text{Rolling period } P = \frac{2\pi k}{\sqrt{g \, GM}} \text{ seconds}$$

where GM is the metacentric height

k is the radius of gyration of the loaded ship about a longitudinal polar axis.

Thus a large metacentric height will produce a small period of roll, although the movement of the ship may be decidedly uncomfortable and possibly dangerous. A small metacentric height will produce a long period of roll and smooth movement of the ship. The resistance to heel, however, will be small and consequently large amplitudes of roll may be experienced. The value of the radius of gyration will vary with the disposition of the cargo. For dry cargo ships, where the cargo is stowed right across the ship, the radius of gyration varies only slightly with the condition of loading and is about 35% of the midship beam.

It is difficult in this type of ship to alter the radius of gyration sufficiently to cause any significant change in the rolling period. Variation in the period due to changes in metacentric height are easier to achieve.

In tankers and OBO vessels it is possible to change the radius of gyration and not as easy to change the metacentric height. If the cargo is concentrated in the centre compartments, with the wing tanks empty, the value of the radius of gyration is small, producing a small period of roll. If, however, the cargo is concentrated in the wing compartments, the radius of gyration increases, producing a slow rolling period. This phenomenon is similar to a skater spinning on ice; as the arms are outstretched the spin is seen to be much slower.

Problems may occur in a ship which travels in a beam sea, if the period of encounter of the waves synchronises with the natural frequency of roll. Even with small wave forces the amplitude of the roll may increase to alarming proportions. In such circumstances it may be necessary to change the ship's heading and alter the period of encounter of the waves.

REDUCTION OF ROLL

Bilge keels
When ships were first built of iron instead of wood a bar keel was fitted, one of its advantages being that it acted as an anti-rolling device. With the fitting of the flat plate keel the anti-rolling properties were lost. An alternative method was supplied in the form of bilge keels which are now used in the majority of ships. These projections are arranged at the bilge to lie above the line of the bottom shell and within the breadth of the ship, thus being partially protected against damage. The depth of the bilge keels depends to some extent on the size of the ship but there are two main factors to be considered;

    (a) the web must be deep enough to penetrate the boundary layer of water travelling with the ship

    (b) if the web is too deep the force of water when rolling may cause damage.

Bilge keels 250 mm to 400 mm in depth are fitted to ocean-going ships. The keels extend for about one half of the length of the ship amidships and are tapered gradually at the ends. Some forms of bilge keel are shown in Fig. 11.10 They are usually fitted in two parts, the connection to the shell plating being stronger than the connection between the two parts. In this way it is more likely, in the event of damage, that the web will be torn from the connecting angle than the connecting angle from the shell.

The bilge keels reduce the initial amplitude of roll as well as subsequent movements.

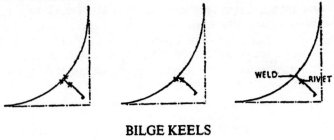

BILGE KEELS

Fig. 11.10

Active fin stabilisers

Two fins extend from the ship side at about bilge level. They are turned in opposite directions as the ship rolls. The forward motion of the ship creates a force on each fin and hence produces a moment opposing the roll. When the fin is turned down, the water exerts an upward force. When the fin is turned up, the water exerts a downward force.

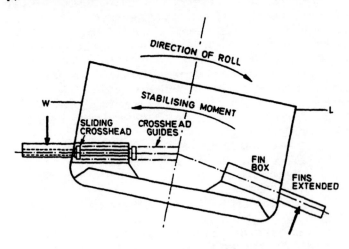

FIN STABILISER

Fig. 11.11

The fins are usually rectangular, having aerofoil cross-section, and turn through about 20°. Many are fitted with tail fins which turn relative to the main fin through a further 10°. The fins are turned by means of an electric motor driving a variable delivery pump, delivering oil under pressure to the fin tilting gear. The oil actuates rams coupled through a lever to the fin shaft.

Most fins are retractable, either sliding into fin boxes transversely or hinged into the ship. Hinged fins are used when there is a restriction on the width of ship-which may be allocated, such as in a container ship.

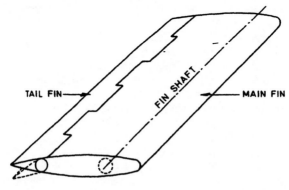

STABILISER FIN   Fig. 11.12

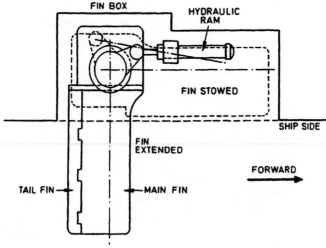

HINGED FIN   Fig. 11.13

The equipment is controlled by means of two gyroscopes, one measuring the angle roll and the other the velocity of roll. The movements of the gyroscopes actuate relays which control the angle and direction through which the fins are turned. It should be noted that no movement of stabiliser can take place until there is an initial roll of the ship and that the fins require a forward movement of the ship to produce a righting moment.

Tank stabilisers

There are three basic systems of roll-damping using free surface tanks:

(a)   Passive Tanks
(b)   Controlled Passive Tanks
(c)   Active Controlled Tanks

These systems do not depend upon the forward movement of the ship and are therefore suitable for vessels such as drill ships. In introducing a free surface to the ship, however, there is a reduction in stability which must be considered when loading the ship.

(a)   Passive Tanks

Two wing tanks are connected by a duct having a system of baffles (Fig. 11.14). The tanks are partly-filled with water. When the ship rolls, the water moves across the system in the direction of the roll. As the ship reaches its maximum angle and commences to return, the water, slowed by the baffles,

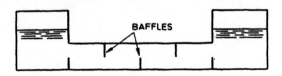

PASSIVE TANK SYSTEM

Fig. 11.14

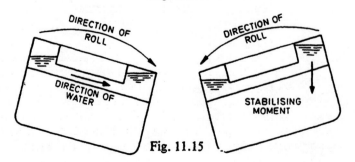

Fig. 11.15

continues to move in the same direction. Thus a moment is created, reducing the momentum of the ship and hence the angle of the subsequent roll. (Fig. 11.15).

The depth of water in the tanks is critical and, for any given ship, depends upon the metacentric height. The tank must be tuned for any loaded condition by adjusting the level, otherwise the movement of the water may synchronise with the roll of the ship and create dangerous rolling conditions. Alternatively the cross-sectional area of the duct may be adjusted by means of a gate valve.

(b) Controlled Passive Tanks (Fig. 11.16)

The principle of action is the same as for the previous system, but the transverse movement of the water is controlled by valves operated by a control system similar to that used in the fin stabiliser. The valves may be used to restrict the flow of water in a U-tube system, or the flow of air in a fully-enclosed system. The mass of water required in the system is about 2% to 2½% of the displacement of the ship.

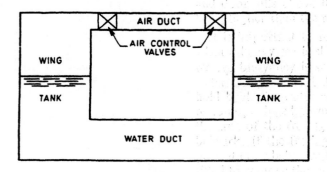

CONTROLLED PASSIVE TANK

Fig. 11.16

(c) Active Controlled Tanks (Fig. 11.17)

In this system the water is positively driven across the ship in opposition to the roll. The direction of roll, and hence the required direction of the water, changes rapidly. It is therefore necessary to use a uni-directional impeller in conjunction with a series of valves. The impeller runs continually and the direction of the water is controlled by valves which are activated by a gyroscope system similar to that used for the fin stabiliser.

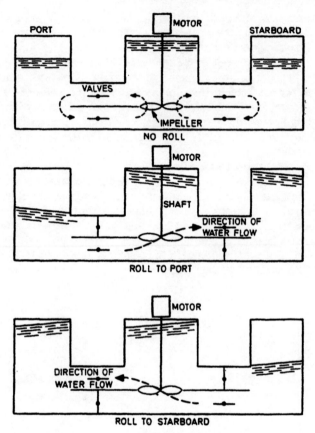

ACTIVE CONTROLLED TANK

Fig. 11.17

VIBRATION

Ship vibration is the periodic movement of the structure and may occur vertically, horizontally or torsionally.

There are several sources of ship vibration, any of which could cause discomfort to personnel, damage to fittings and instruments and structural failure. If the frequency of the main or auxiliary machinery at any given speed coincides with the natural frequency of the hull structure, then vibration may occur. In such circumstances it is usually easier to alter the source of the vibration by changing the engine speed or fitting dampers, than to change the structure. The natural frequency of the structure depends upon the length, mass distribution and second moment of area of the structural material. For any given mass distribution a considerable change in structural material would be required to cause any practical variation in natural frequency. There is a possibility of altering the natural frequency of the hull by re-distributing the cargo. If the cargo is concentrated at the nodes, the natural frequency will be increased. If the cargo is concentrated at the anti-nodes, the natural frequency and the deflection will be reduced. Such changes in cargo distribution may only be possible in vessels such as oil tankers or bulk carriers in the ballast condition.

Similarly vibration may occur in a machinery space due to unbalanced forces from the main or auxiliary machinery or as a result of uneven power distribution in the main engine. This vibration may be transmitted through the main structure to the superstructure, causing extreme discomfort to the personnel.

As explained earlier the variation in blade loading due to the wake may create vibration of the after end, which may be reduced by changing the number of blades. The turbulence of the water caused by the shape of the after end is also a source of vibration which may be severe. It is possible to design the after end of the ship to reduce this turbulence resulting in a smoother flow of water into the propeller disc.

Severe vibration of the after end of some ships has been caused by insufficient propeller tip clearance. As the blade tip passes the top of the aperture it attempts to compress the water. This creates a force on the blade which causes bending of the blade and increased torque in the shaft. The periodic nature of this force, i.e., revs × number of blades, produces the vibration of the stern. Classification Societies recommend minimum tip clearances to reduce propeller-induced vibrations to reasonable levels. Should the tip clearance be constant, e.g., with a propeller nozzle, then this problem does not occur. If an existing ship suffers an unacceptable level of vibration from this source, it may be necessary to crop the blade tips, reducing the propeller efficiency, or to fit a propeller of smaller diameter.

A damaged propeller blade will create out-of-balance moments due to the unequal weight distribution and the uneven loading on the blades. Little may be done to relieve the resultant vibration except to repair or replace the propeller.

Wave induced vibrations may occur in ships due to pitching, heaving, slamming or the passage of waves along the ship. In smaller vessels pitching and slamming are the main sources but are soon dampened. In ships over about 300 m in length, however, hull vibration has been experienced in relatively mild sea states due to the waves. In some cases the vibration has been caused by the periodic increase and decrease in buoyancy with regular waves much shorter than the ship, whilst in other cases, with non-uniform waves, the internal energy of the wave is considered to be the source. Such vibrations are dampened by a combination of the hull structure, the cargo, the water friction and the generation of waves by the ship.

CHAPTER 12

MISCELLANEOUS

INSULATION OF SHIPS

Steelwork is a good conductor of heat and is therefore said to have a high thermal conductivity. It will therefore be appreciated that some form of insulation, having low thermal conductivity, must be fitted to the inner face of the steelwork in refrigerated compartments to reduce the transfer of heat. The ideal form of insulation is a vacuum although good results may be obtained using air pockets. Most insulants are composed of materials having entrapped air cells, such as cork, glass fibre and foam plastic. Cork may be supplied in slab or granulated form, glass fibre in slab form or as loose fill, while foam plastic may either be supplied as slabs or the plastic may be foamed into position. Granulated cork and loose glass fibre depend to a large extent for their efficiency on the labour force, while in service they tend to settle, leaving voids at the top of the compartment. These voids allow increased heat transfer and plugs are fitted to allow the spaces to be repacked. Horizontal stoppers are arranged to reduce settlement. Glass fibre has the advantages of being fire resistant, vermin-proof and will not absorb moisture. Since it is also lighter than most other insulants it has proved very popular in modern vessels. Foam plastic has recently been introduced, however, and when foamed in position, entirely fills the cavity even if it is of awkward shape.

The depth of insulation in any compartment depends upon the temperature required to maintain the cargo in good condition, the insulating material and the depth to which any part of the structure penetrates the insulant. It is usually found that the depths of frames and beams govern the thickness of insulation at the shell and decks, 25 mm to 50 mm of insulation projecting past the toe of the section. It therefore proves economical in insulated ships to use frame and reverse as shown in Fig. 3.7 and thus reduce the depth of insulation. This also also has the effect of increasing the cargo capacity.

The internal linings required to retain and protect the insulation may be of galvanised iron, stainless steel or aluminium alloy. The linings are screwed to timber grounds which are, in turn, connected to the steel structure. The linings are made airtight by coating the overlaps with a composition such as white lead and fitting sealing strips. This prevents heat transfer due to a circulation of air and prevents moisture entering the insulation. Cargo battens are fitted to all the exposed surfaces to prevent contact between the cargo and the linings and to improve the circulation of air round the cargo. Fig. 12.1 shows a typical arrangement of insulation at the ship side.

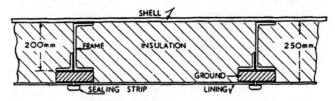

SHELL INSULATION

Fig. 12.1

At the tank top an additional difficulty arises, that of providing support for the cargo. Much depends upon the load bearing qualities of the insulant. Slab cork, for instance, is much superior in this respect to glass fibre, and may be expected to carry some part of the cargo load. The tank top arrangement in such a case would be as shown in Fig. 12.2.

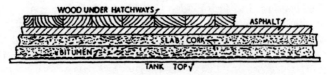

CORK INSULATION OF TANK TOP

Fig. 12.2

Slab cork 150 mm to 200 mm thick is laid on hot bitumen. The material is protected on its upper surface by a layer of asphalt 50 mm thick which is reinforced by steel mesh. Wood ceiling is fitted under the hatchways where damage is most likely to occur.

Fig. 12.3 shows the arrangement where glass fibre is used. The cargo is supported on double timber ceiling which is supported by bearers about 0.5 m apart. An additional thickness of ceiling is fitted under the hatchways. The glass fibre is packed between the bearers. If oil fuel is carried in the double bottom in way of an insulated space it is usual to leave an air gap between the tank top and the insulation. This ensures that any leakage of oil will not affect the insulation.

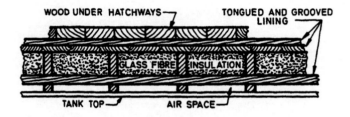

GLASS FIBRE INSULATION OF TANK TOP

Fig. 12.3

Particular care must be taken to design the hatchways to avoid heat transfer. The normal hatch beams are fitted with tapered wood which is covered with galvanised sheet. A similar type of arrangement is made at the ends of the hatch. Insulated plugs with opposing taper are wedged into the spaces. The normal hatch boards are fitted at the top of the hatch as shown in Fig. 12.4. Steel hatch covers may be filled with some suitable insulation and do not then require separate plugs.

Similar types of plug are fitted at the bilge to allow inspection and maintenance, and in way of tank top manholes.

Drainage of insulated spaces is rather difficult. Normal forms of scupper would lead to increase in temperature. It becomes necessary, therefore, to fit brine traps to all 'tween deck and hold spaces. After defrosting the compartments and removing the cargo the traps must be re-filled with saturated brine, thus forming an air seal which will not freeze. Fig. 12.5 shows a typical brine trap fitted in the 'tween decks.

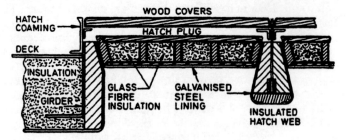

INSULATED HATCH PLUG

Fig. 12.4

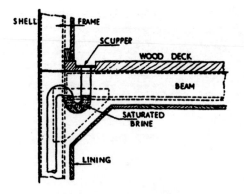

TWEEN DECK SCUPPER

Fig. 12.5

CORROSION

Corrosion is the wasting away of a material due to its tendency to return to its natural state, which, in the case of a metal, is in the form of an oxide.

If a metal or alloy is left exposed to a damp atmosphere, an oxide will form on the surface. If this layer is insoluble, it forms a protective layer which prevents any further corrosion. Copper and aluminium are two such metals. If, on the other hand, the layer is soluble, as in the case of iron, the oxidation continues, together with the erosion of the material.

When two dissimilar metals or alloys are immersed in an electrolyte, an electric current flows through the liquid from one metal to the other and back through the metals. The direction in which the current flows depends upon the relative position of the metals in the electrochemicals series. For common metals in use in ships, this series is in the following order of electrode potential:

copper   +
lead
tin
iron
chromium
zinc
aluminium
magnesium   –

The current flows from the anode to the cathode which is higher in the scale, or more electro-positive. Thus, if copper and iron are joined together and immersed in an electrolyte, a current will flow through the electrolyte from the iron to the copper and back through the copper to the iron (Fig. 12.6).

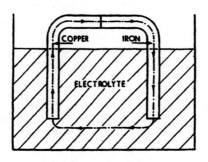

CORROSION CELL

Fig. 12.6

Unfortunately, during this process, material from the anode is transferred to the cathode, resulting in corrosion of the anode. Slight differences in potential occur in the same material. Steel plate, for instance, is not perfectly homogeneous and will therefore have anodic and cathodic areas. Corrosion may therefore occur when such a plate is immersed in an electrolyte such as sea water. The majority of the corrosion of ships is due to this electrolytic process.

PREVENTION OF HULL CORROSION

Corrosion may be prevented by coating the material with a substance which prevents contact with moisture or the electrolyte. In principle this is simple to achieve but in practice it is found difficult to maintain such a coating, particularly on ships which may be slightly damaged by floating debris and rubbing against quays.

Surface preparation.
Steel plates supplied to shipbuilders have patches of a black oxide known as *mill scale* adhering to the surface. This scale is insoluble and, if maintained over the whole surface, would reduce corrosion. It is, however, very brittle and does not expand either mechanically or thermally at the same rate as the steel plate. Unless this mill scale is removed before painting, the painted scale will drop off in service, leaving bare steel plate which will corrode rapidly. Unfortunately mill scale is difficult to remove completely.

If the plate is left exposed to the atmosphere, rust will form behind the mill scale. On wire brushing, the majority of the scale will be removed. This is known as *weathering*. In modern times a good flow of material through the shipyard is essential and therefore the time allowed for weathering must be severely limited. In addition, it is found in practice that much of the mill scale is not removed by this process.

If the plates are immersed in a weak solution of sulphuric acid or hydrochloric acid for a few hours, the majority of the mill scale is removed. This system, known as *pickling,* has been used by the Admiralty and several private owners for many years. The pickled plate must be hosed down with fresh water on removal from the tank, to remove all traces of the acid. It is then allowed to dry before painting. One disadvantage with this method is that during the drying period a light coating of rust is formed on the plate and must be removed before painting.

Flame cleaning of ship structure came into use some time ago. An oxy-acetylene torch, having several jets, is used to brush the surface. It burns any dirt and grease, loosens the surface rust and, due to the differential expansion between the steel and the mill scale, loosens the latter. The surface is immediately wire brushed and the priming coat applied while the plate is still warm. Opinions vary as to the efficiency of flame cleaning, some shipowners having had excellent results, but it has lost favour in the last few years.

The most effective method of removing the mill scale which has been found to date, is the use of *shot blasting*. The steel plates are passed through a machine in which steel shot is projected at the plate, removing the mill scale together with any surface rust, dirt and grease. This system removes 95% to 100% of the mill scale and results in a slightly rough surface which allows adequate adhesion of the paint. In modern installations, the plate is spray painted on emerging from the shot blasting machine.

Painting

Work done in efficient surface preparation is wasted unless backed up by suitable paint correctly applied. The priming coat is perhaps the most important. This coat must adhere to the plate and, if applied before construction, must be capable of withstanding the wear and tear of everyday working. The subsequent coats must form a hard wearing, watertight cover. The coats of paint must be applied on clean, dry surfaces to be completely effective.

Cathodic protection

If three dissimilar metals are immersed in an electrolyte, the metal lowest on the electro chemical scale becomes the anode, the remaining two being cathodes. Thus if copper and iron are immersed in sea water, they may be protected by a block of zinc which is then known as a *sacrificial anode,* since it is allowed to corrode in preference to the copper and iron. Thus zinc or magnesium anodes may be used to protect the propeller and stern frame assembly of a ship, and will, at the same time, reduce corrosion of the hull due to differences in the steel.

Deep water ballast tanks may be protected by sacrificial anodes. It is first essential to remove any rust or scale from the surface and to form a film on the plates which prevents any further corrosion. Both of these functions are performed by booster anodes which have large surface area compared with their volume (*e.g.,* flat discs). These anodes allow swift movement of material to the cathode, thus forming the film. Unfortunately this film is easily removed in service and therefore main anodes are fitted, having large volume compared with surface area (*e.g.,* hemispherical), which are designed to last about three years. Protection is only afforded to the whole tank if the electrolyte is in contact with the whole tank. Thus it is necessary when carrying water ballast to press the tank up.

Electrolytic action may occur when two dissimilar metals are in contact above the waterline. Great care must be taken, for instance, when joining an aluminium alloy deckhouse to a steel deck. Several methods have been tried with varying degrees of success. The steel bar forming the attachment may be galvanised, steel or iron rivets being used through the steel deck, with aluminium rivets to the deckhouse. A coating of barium chromate between the surfaces forms a measure of protection.

The method used on M.S. *Bergensfjord* was most effective, although perhaps costly. Contact between the two materials was prevented by fitting 'Neoprene' tape in the joint (Fig. 12.7). Galvanised steel bolts were used, but 'Neoprene' ferrules were fitted in the bolt holes, opening out to form a washer at the bolt head. The nut was fitted in the inside of the house, and tack welded to the boundary angle to allow the joint to be tightened without removal of the internal lining. The top and bottom of the joints were then filled with a compound known as 'Aranbee' to form a watertight seal.

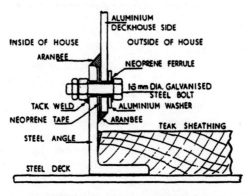

CONNECTION OF ALUMINIUM
DECKHOUSE TO STEEL DECK

Fig. 12.7

Impressed current system

A more sophisticated method of corrosion control of the outer shell may be achieved by the use of an impressed current. A number of zinc reference anodes are fitted to the hull but insulated from it. It is found that the potential difference between the anode and a fully protected steel hull is about 250 mV. If the measured difference at the electrode exceeds this value, an electric current is passed through a number of long lead-silver alloy anodes attached to, but insulated from, the hull. The protection afforded is more positive than with sacrificial anodes and it is found that the lead-silver anodes do not erode. A current of 7 to 350 mA/m² is required depending upon the surface protection and the degree to which the protection has broken down.

Design and maintenance

Corrosion of ships may be considerably reduced if careful attention is paid to the design of the structure. Smooth, clean surfaces are easy to maintain and therefore welded ships are preferable from this point of view. Riveted seams and stiffeners tend to harbour moisture and thus encourage corrosion. If parts of the structure are difficult to inspect, then it is unlikely that these parts will be properly maintained. Efficient drainage of all compartments should be ensured.

Those parts of the structure which are most liable to corrosion should be heavily coated with some suitable compound if inspection is difficult. Steel plating under wood decks or deck composition is particularly susceptible in this respect. Pools of water lying in plate edges on the deck tend to promote corrosion. If such pools may not be avoided then the plate edges must be regularly painted. Such difficulties arise with joggled deck plating.

A warm, damp atmosphere encourages corrosion. Care must be taken, therefore, to regularly maintain the structure in way of deck steam pipes and galley funnels.

Reductions in thickness of material of between 5% and 10% may be allowed if the structure is suitably protected against corrosion. If an impressed current system is used for the hull, the maximum interval between dockings may be increased from 2 to 2½ years.

FOULING

The resistance exerted by the water on a ship will be considerably increased if the hull is badly fouled by marine growth. It is found that marine growth will adhere to the ship if the speed is less than about 4 knots. Once attached, however, the growth will continue and will be difficult to remove despite the speed. The type of fouling depends upon the nature of the plant and animal life in the water.

It is essential to reduce fouling, since the increase in resistance in severe cases may be in the order of 30% to 40%. This is reflected in an increase in fuel consumption to maintain the same speed, or a reduction in speed for the same power.

The main anti-fouling system is the use of toxic coatings — usually mercury based. The coating exudes a poison which inhibits the marine growth. Unfortunately the poison works at all times. Thus when the ship speed exceeds 4 knots, poison is being wasted. After a period the outer layer of the coating is devoid of mercury and the remainder is unable to work its way to the water. If the shell is scrubbed at this stage, the outer dead layer is removed and the coating once again becomes active. The scrubbing may be carried out using nylon brushes, either in dock or while afloat.

A recent innovation in the anti-fouling campaign is the introduction of self-polishing copolymers (spc). This is a paint in which the binder and toxins are chemically combined. Water in contact with the hull causes a chemical and physical change on the surface of the coating. When water flows across the surface, the local turbulence removes or *polishes* this top layer. This produces a controlled rate of release of toxin, the life of which depends upon the thickness of the coating. Most coatings are designed to last at least 24 months.

In addition to the anti-fouling properties, the polishing of the layers produces a very smooth surface and hence considerably reduces the frictional resistance and hence the fuel consumption. This is in addition to the reduction in resistance due to the reduced fouling.

Fouling of the sea inlets may cause problems in engine cooling, whilst explosions have been caused by such growths in air pipes.

When a ship is in graving dock, hull fouling must be removed by scrapers or high pressure water jets. These water jets, with the addition of abrasives such as grit, prove very effective in removing marine growth and may be used whilst the vessel is afloat.

One method of removing growth is by means of explosive cord. The cord is formed into a diamond-shaped mesh which is hung down from the ship side, attached to a floating line. The cord is energised by means of a controlled electrochemical impulse. The resultant explosion produces pressure waves which pass along the hull, sweeping it clear of marine growth and loose paint. By energising the net in sequential layers, the hull is cleaned quickly but without the excessive energy which would result from a single charge.

EXAMINATION IN DRY DOCK

In many companies it is the responsibility of the marine engineers to inspect the hull of the ship on entering graving dock, while in other companies it is the responsibility of the deck officers. It is essential on such occasions to make a thorough examination to ensure that all necessary work is carried out.

The shell plating should be hosed with fresh water and brushed down immediately to remove the salt before the sea water dries. The plating must be carefully checked for distortion, buckling, roughness, corrosion and slack rivets. In the case of welded ships the butts and seams should be inspected for cracks. The side shell may be slightly damaged due to rubbing against quays. After inspection and repair the plating should be wire brushed and painted. Any sacrificial anodes must be checked and replaced if necessary, taking care not to paint over the surface. The ship side valves and cocks are examined, glands repacked and greased. All external grids are examined for corrosion and freed from any blockage. If severe wastage has occurred the grid may be built up with welding. The shell boxes are wire brushed and painted with an anti-fouling composition.

If the double bottom tanks are to be cleaned, the tanks are drained by unscrewing the plugs fitted at the after end of the tank. This allows complete drainage since the ship lies at a slight trim by the stern. It is essential that these plugs should be replaced before undocking, new grummets always being fitted.

The after end must be examined with particular care. If at any time the ship has grounded, the sternframe may be damaged. It should be carefully inspected for cracks, paying particular attention to the sole piece. In twin screw ships the spectacle frame must be thoroughly examined. The drain plug in the bottom of the rudder is removed to determine whether any water has entered the rudder. Corrosion on the external surface may be the result of complete wastage of the plate from the inside. The weardown of the rudder is measured either at the tiller or at the upper gudgeon. Little or no weardown should be seen if the rudder is supported by a carrier, but if there are measurable differences the bearing surfaces of the carrier should be examined. If no carrier is fitted, appreciable weardown may necessitate replacing the hard steel bearing pad in the lower gudgeon. The bearing material in the gudgeons must be examined to see that the pintles are not too slack, a clearance of 5 mm being regarded as a maximum. The pintle nuts, together with any form of locking device, must be checked to ensure that they are tight.

Careful examination of the propeller is essential. Pitting may occur near the tips on the driving face and on the whole of the fore side due to cavitation. Propeller blades are sometimes damaged by floating debris which is drawn into the propeller stream. Such damage must be made good as it reduces the propeller efficiency, while the performance is improved by polishing the blade surface. If a built propeller is fitted, it is necessary to ensure that the blades are tight and the pitch should be checked at the same time. The stern gland should be carefully repacked and the propeller nut examined for movement. Such movement usually results in cracking of the cement filling which is readily seen. The weardown of the tailshaft should be measured by inserting a wedge between the shaft and the packing. If this weardown exceeds about 8 mm the bearing material should be renewed, 10 mm being regarded as an absolute maximum. There should be little or no weardown in an oil lubricated stern tube. The weardown in this type is usually measured by means of a special gauge as the sealing ring does not allow the insertion of a weardown wedge.

The efficiency and safety of the ship depend to a great extent on the care taken in carrying out such an inspection.

EMERGENCY REPAIRS TO STRUCTURE

During a ship's life faults may occur in the structure. Some of these faults are of little importance and are inconvenient rather than dangerous. Other faults, although apparently small, may be the source of major damage.

It is essential that a guided judgement be made of the relative importance of the fault before undertaking repairs.

In Chapter 2 it was explained that the highest longitudinal bending moments usually occur in the section of the ship between about 25% forward and aft of midships. Continuous longitudinal material is provided to maintain the stresses at an acceptable level. If there is a serious reduction in cross-sectional area of this material, then the ship could split in two. Damage to the plating at the fore end or the stern is of less importance, although flooding could occur, whilst a crack in a rudder plate is inconvenient.

If a plate is damaged, several options are available
i)   Replace the plate
ii) Cut out and weld any cracks
iii) Fill in any pits with welding
iv) Fit a welded patch over the fault

v) Stop a leak by means of a cement box

vi) Cut off any loose plating.

The ultimate solution in all cases is to replace the plate, but in an emergency or on a short term basis, the other options must be considered.

If a crack occurs in a plate, then a hole should be drilled at each end of the crack to prevent its propagation. If the plate is in the midship part of the ship, then great care must be taken. Any welding must be carried out by authorised personnel in the presence of a Classification Society surveyor. The correct weld preparation and welding sequence must be used. If the plating is of high tensile steel the correct welding rods must be used. Greater damage may be caused by an untrained welder than leaving the crack untreated.

Pitting in a plate may be filled up with weld metal except in the midship region. In this section it is better to clean out the pit, grind the surface smooth and fair with some suitable filling material to prevent an accumulation of water.

A crack in a rudder plate may be patched once the crack has been stopped. Water in a rudder may increase the load on the rudder carrier and steering gear but has little other effect.

Damage to the fore end usually results in distortion of the structure and leakage. It may be possible to partially remove the distortion with the aid of hydraulic jacks, in which case the plating may be patched. Otherwise it is probably necessary to fit a cement box.

Damage to a bilge keel may prove serious. In this case it is better to cut off any loose material and to taper the material on each side of the damage, taking care to buff off any projecting material or welding in way of the damage. A replacement bilge keel must be fitted in the presence of a Classification Society Surveyor.

ENGINE CASING

The main part of the machinery space in a ship lies between the double bottom and the lowest deck. Above this deck is a large vertical trunk known as the engine casing, which extends to the weather deck. In the majority of ships this casing is surrounded by accommodation. An access door is fitted in each side of the casing, leading into the accommodation. This door may be of wood unless fitted in exposed casings. At the top of the trunk the funnel and engine room skylight are fitted. The skylight supplies natural light to the engine room and may be opened to supplement the ventilation, the whole casing then acting as an air trunk.

The volume taken up by the casings is kept as small as possible since, apart from the light and air space, they serve no useful purpose. It is essential, however, that the minimum width and length should be sufficient to allow for removal of the machinery. The dimensions of the engine room skylight are determined on the same basis and it is common practice to bolt the skylight to the deck to facilitate removal.

The casings are constructed of relatively thin plating with small vertical angle stiffeners about 750 mm apart. In welded ships, flat bars may be used or the plating may be swedged. The stiffeners are fitted inside the casing and are therefore continuous. Pillars or deep cantilevers are fitted to support the casing sides. Cantilevers are fitted in many ships to dispense with the pillars which interfere with the layout of machinery. The cantilevers are fitted in line with the web frames.

The casing sides in way of accommodation are insulated to reduce the heat transfer from the engine room. While such transfer would be an advantage in reducing the engine room temperature, the accommodation would be most uncomfortable. A suitable insulant would be glass fibre in slab form since it has high thermal efficiency and is fire resistant. The insulaton is fitted inside the casing and is faced with sheet steel or stiffened cement.

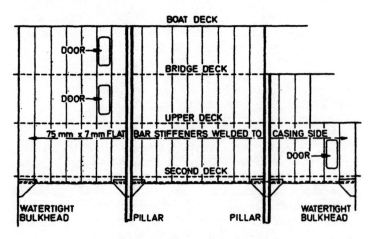

ELEVATION OF ENGINE CASING

Fig. 12.8

Two strong beams are fitted at upper deck or bridge deck level to tie the two sides of the casing together. These strong beams are fitted in line with the web frames and are each in the form of two channel bars at adjacent frame spaces, with a shelf plate joining the two channels.

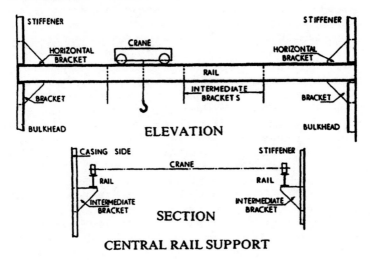

CENTRAL RAIL SUPPORT

Fig. 12.9

Efficient lifting gear is essential in the engine room to allow the removal of machinery parts for inspection, maintenance and repair. The main equipment is a travelling crane of 5 or 6 tonne lifting capacity on two longitudinal rails which run the full length of the casing. The rails are formed by rolled steel joists, efficiently connected to the ends of the casing or the engine room bulkheads by means of large brackets. Intermediate brackets may be fitted to reduce movement of the rails, as long as they do not obstruct the crane.

Fig. 12.9 shows a typical arrangement.

The height of the rails depends upon the height and type of the machinery, sufficient clearance being allowed to remove long components such as pistons and cylinder liners. The width between the rails is arranged to allow the machinery to be removed from the ship.

An alternative method used in some ships is to carry two cranes on transverse rails. This reduces the length of the rails but no intermediate support may be fitted.

FUNNEL

The size and shape of the funnel depends upon the requirements of the shipowners. At one time tall funnels were fitted to steam ships to obtain the required natural draught and, in passenger ships, to ensure that the smoke and grit were carried clear of the decks. Modern ships with forced draught do not require such high funnels. The funnel has now become a feature of the design of the ship, enhancing the appearance and being a suitable support for the owners' housemark. They are built much larger than necessary, particularly in motor ships where the uptakes are small. They may be circular, elliptical or pear shaped in plan view, while there are many varied shapes in side elevation. In many cases the funnel is designed to house deck stores or auxiliary machinery such as ventilating fan units.

The funnel consists of an outer casing protecting the uptakes. The outer funnel is constructed of steel plate 6 mm to 8 mm in thickness. It is stiffened internally by ordinary angles or flat bars fitted vertically. Their scantlings depend upon the size and shape of the funnel. The plating is connected to the deck by a boundary angle, while a moulding is fitted round the top to stiffen the free edge. Steel wire stays are connected to lugs on the outside of the funnel and to similar lugs on the deck. A rigging screw is fitted to each stay to enable the stays to be tightened. A watertight door is fitted in the funnel, having clips which may be operated from both sides (Fig. 12.10).

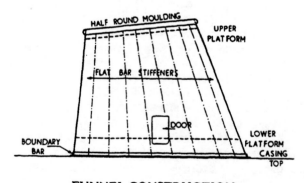

FUNNEL CONSTRUCTION

Fig. 12.10

The uptakes from the boilers, generators and main engine are carried up inside the funnel and stopped almost level with the top of the funnel (Fig. 12.11). A steel platform is fitted at a height of about 1 m inside the funnel. This platform extends right across the funnel, holes being cut in for the uptakes and access. The uptakes are not connected directly to this platform because of possible expansion, but a ring is fitted above and below the plating, with a gap which allows the pipe to slide. Additional bellows expansion joints are arranged where necessary. At the top a single platform or separate platforms may be fitted to support the uptakes, the latter being connected by means of an angle ring to the platform. In motor ships a silencer must be fitted in the funnel to the main engine exhaust. This unit is supported on a separate seat. Ladders and gratings are fitted inside the funnel to allow access for inspection and maintenance.

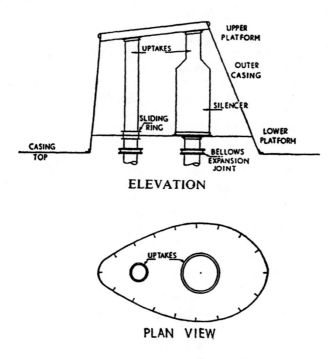

ELEVATION

PLAN VIEW

ARRANGEMENT OF FUNNEL UPTAKES

Fig. 12.11

SELECTION OF EXAMINATION QUESTIONS

CLASS 2

1. Sketch and describe a watertight door. What routine maintenance must be carried out to ensure that the door is always in working order?

2. Draw an outline midship section of a ship and show the position of the following items: (a) sheer strake, (b) garboard strake, (c) stringer, (d) bilge plating, (e) keel plate, (f) floors, (g) frames.

3. With the aid of a sketch showing only the compartments concerned, show the arrangement of windlass and anchor cables. How is the end of the cable secured in the chain locker? What is meant by the terms: (a) hawse pipe, (b) spurling pipe, (c) cable lifter, (d) cable stopper?

4. Sketch and describe the construction of a cruiser stern fitted to a single screw ship and discuss its advantages.

*5. Show, with the aid of diagrammatic sketches, how a large ship is supported in dry dock. Indicate the strains that are imposed on a ship resting on the blocks. Detail the precautions that should be taken when refloating the ship in a dry dock.

6. Sketch and describe a transverse section of either an oil tanker or an ore carrier having two longitudinal bulkheads.

*7. Describe with sketches a cargo hatch fitted with steel covers and show how the structural strength of the deck is maintained. Compare the relative merits of steel and wooden hatch covers.

*Questions marked with an asterisk have been selected from Department of Transport papers and are reproduced by kind permission of The Controller of Her Majesty's Stationery Office.

8. Explain what is meant by *longitudinal framing* and *transverse framing*. Which types of ships would have these methods of construction?

9. Explain with the aid of sketches the terms *hogging* and *sagging* with reference to a ship meeting waves having the same length as the vessel. What portions of the structure resist these stresses?

10. Describe with the aid of sketches the terms: (a) camber, (b) sheer, (c) rise of floor, (d) flare. What is the purpose of these?

11. A ship has a small hole below the waterline. What would be the procedure in making a cement box round the hole?

12. Show how the hatch and main hold of a refrigerated vessel are insulated. What materials are used? How are the compartments drained?

13. Sketch and describe a stern frame. Show how the frame is attached to the adjoining structure. State the materials used together with their properties.

14. Sketch and describe a deep tank, giving details of the watertight hatch.

15. Sketch and describe a weather deck hatch coaming giving details of the attachment of the half beams. Do the half beams give any strength to the deck?

16. Sketch and describe the different floors used in the construction of a double bottom, indicating where each type is employed. Give details of the attachment of the floors to the adjacent structure.

17. Sketch and describe a transverse watertight bulkhead of a cargo vessel. Show details of the stiffening and the boundary connections.

*18. Describe with the aid of a diagrammatic sketch the following types of keel. Show how they are attached to the ship's hull and indicate on the sketch the main structural members: (a) bilge keel, (b) flat plate keel, (c) duct keel.

19. What are the main functions of: (a) fore peak, (b) after peak, (c) deep tank, (d) double bottom? Give examples of the liquids carried in these tanks.

20. Describe the causes of corrosion in a ship's structure and the methods used to reduce wastage. What parts of the ship are most liable to attack?

21. Sketch and describe a rudder suitable for a ship 120 m long and speed 14 knots. Show how the rudder is supported.

*22. Enumerate the examination and tests which should be carried out on the exterior of a ship's hull when in dry dock. Detail the inspection necessary in the region of the ship's hull which is adjacent to the main machinery spaces. Discuss the nature of the defects liable to be found in these areas.

*23. Explain, with the aid of diagrammatic sketches where applicable, the meaning of the following terms: (a) spurling pipe, (b) centre girder, (c) cofferdam, (d) collision bulkhead, (e) metacentre.

24. What precautions must be taken on entering ballast or fuel tanks when empty? Explain why these precautions are necessary.

*25. With reference to ships transporting liquefied petroleum gases (LPG) describe TWO of the following methods of carriage: (a) pressurised system, (b) fully refrigerated system, (c) semi refrigerated system.

26. Sketch and describe the spectacle frame of a twin screw ship. Show how it is attached to the ship.

27. Explain why the plating of the hull and transverse watertight bulkheads are arranged horizontally. Which sections of the ship's structure constitute the strength members, and what design considerations do they receive?

28. Define the following terms: (a) displacement, (b) gross tonnage, (c) net tonnage, (d) deadweight.

29. What are cofferdams and where are they situated? Describe their use in oil tankers.

30. Sketch the panting arrangements at the fore end of a vessel.

*31. With respect to stability explain what is meant by: (a) stiff, (b) tender. Comment on the type of loading or cargo associated with each of the above conditions.

32. Explain why and where transverse bulkheads are fitted in a ship. In which ships are longitudinal bulkheads fitted and what is their purpose?

*33. Sketch the construction of a bulbous bow and briefly comment on the advantages of fitting this bow to certain vessels.

34. Sketch and describe an arrangement of funnel uptakes for a motor ship or a steam ship, giving details of the method of support in the outer funnel.

*35. Give a reason for corrosion in each of the following instances:
   (i) Connection between aluminium superstructure and steel deck.   (3 marks)
   (ii) In crude oil cargo tanks.   (3 marks)
   (iii) Explain how in each case corrosion can be inhibited.
                                                       (4 marks)

*36. (i) Sketch in diagrammatic form any type of transverse thrust unit including the power unit, labelling the principal components.   (4 marks)
   (ii) Explain how this unit operates.   (4 marks)
   (iii) Give reasons for its installation in both long-haul container carriers and short sea trade 'ro-ro' ferries.
                                                       (2 marks)

*37. (i) Sketch in diagrammatic form a stabiliser unit in which the fins retract athwartships into a recess in the hull.   (4 marks)
   (ii) Describe how the extension/retraction sequence is carried out.   (4 marks)
   (iii) Define how the action of fin stabilisers effects steering.   (2 marks)

*38. (i) Identify three different corrosion problems encountered in ship structure.   (3 marks)
(ii) Define the origins and significance of each.
    (3 marks)
(iii) State what precautions are taken to reduce their effects.   (4 marks)

*39. Make a sketch of a watertight door giving details of the fastening arrangements to show how edge watertightness is maintained.   (6 marks)
Describe the procedure adopted for testing TWO of the following for watertightness:
(i) A watertight door.   (2 marks)
(ii) A deep tank bulkhead, and   (2 marks)
(iii) A hold-bulkhead in a dry cargo ship.   (2 marks)

*40. Suggest with reasons whether 'build up' by welding, patching, cropping or plate replacement is best suited to the following structural defects:
(i) Severe pitting at one spot on deck stringer. (3 marks)
(ii) External wastage of side plating below scuppers.
    (3 marks)
(iii) Extensive wastage of side plating along waterline.
    (4 marks)

CLASS 1

1. Sketch and describe the methods used to connect the shell plating to the side frames. How is a deck made watertight where pierced by a side frame?

2. Sketch and describe a hatch fitted to an oil tanker. When is this hatch opened?

3. Discuss the forces acting on a ship when floating and when in dry dock.

*4. Enumerate the THREE principal sources of shipboard vibration. Severe vibration is sometimes experienced on vessels in service. Explain how you would trace the source of the forces causing these vibrations and the measures that could be taken to reduce the severity of the vibrations.

*5. Give reasons why a tanker is normally assigned a minimum basic freeboard less than that of other types of ship. Enumerate any supplementary conditions of assignment applicable to tankers.

6. Sketch and describe a gravity davit. What maintenance is required to ensure its efficient working?

7. Sketch a transverse section through a cargo ship showing the arrangement of the frames and the double bottom.

8. Sketch and describe the arrangements to support stern tubes in a twin screw ship.

9. Explain in detail the forces acting on the fore end of a vessel. Sketch the arrangements made to withstand their effects.

*Questions marked with an asterisk have been selected from Department of Transport papers and are reproduced by kind permission of The Controller of Her Majesty's Stationery Office.

10. Draw a longitudinal section of a dry cargo ship with engines amidships with particular reference to the double bottom, showing the spacing of the floors. What types of ships have no double bottom?

11. Sketch and describe a welded watertight bulkhead indicating plate thickness and stiffener sizes. How is it made watertight? Show how ballast pipes, electric cables and intermediate shafting are taken through the bulkhead.

12. Sketch and describe the arrangement of a rudder stock and gland and the method of suspension of a pintleless rudder. How is the wear-down measured? What prevents the rudder jumping?

13. By what means is the fire risk in passenger accommodation reduced to a minimum? Describe with the aid of diagrammatic sketches the arrangements for fire fighting in the accommodation of a large passenger liner.

*14. Describe with the aid of diagrammatic sketches a keyless propeller showing how it is fitted to the tailshaft and discuss the advantages of this design. Give details of the method of driving the propeller on to the shaft and how it is locked in position.

15. Explain what is meant by the following terms: (a) exempted space, (b) deductible space, and (c) net tonnage. Give two examples of (a) and (b).

16. Explain the purposes of a collision bulkhead. Describe with the aid of sketches the construction of a collision bulkhead paying particular attention to the strength and attachment to the adjacent structure.

*17. Describe the precautions and procedures which should be taken to minimise the dangers due to water accumulation during fire fighting: (a) while the ship is in dry dock, (b) while the ship is at sea.

18. What precautions are taken before dry-docking a vessel? What precautions are taken before re-flooding the dock? What fire precautions are taken while in dock?

19. Why and where are deep tanks fitted in cargo ships? Describe the arrangements for filling, emptying and drainage.

*20. Discuss with the aid of diagrammatic sketches THREE of the following ship stabilisation systems: (a) bilge keel, (b) activated fins, (c) active tanks, (d) passive tank.

*21. Aluminium and high tensile steel now often replace mild steel in ship construction. State the types of vessel in which these materials are used and give reasoned explanations why they have replaced mild steel. State the precautions which must be observed when aluminium structures are fastened to steel hulls.

22. Explain with the aid of sketches what is meant by breast hooks and panting beams, giving approximate scantlings. Where are they fitted and what is their purpose?

23. Sketch and describe the freeboard markings on a ship. By what means are they determined? How do the authorities prevent these marks being changed?

24. Sketch and describe the construction of a bulbous bow. Why is such an arrangement fitted?

*25. Describe in detail the main causes of corrosion in a ship's internal structure and the measures which can be taken to minimise this action.
Enumerate the parts of a vessel's internal structure most liable to corrosion.

26. Draw a cross section through a modern oil tanker in way of an oiltight bulkhead.

*27. Enumerate TWO methods of generating inert gas for use in a Very Large Crude Carrier (VLCC).
Describe with the aid of a diagrammatic sketch the distribution layout of the piping and associated equipment of an inert gas system suitable for a VLCC.
Itemise the safety features incorporated in the system.

28. Sketch and describe the construction of a corrugated bulkhead. What are the advantages and disadvantages of such a bulkhead compared with the normal flat bulkhead?

29. Where do discontinuities occur in the structure of large vessels and how are their effects minimised?

30. Sketch and describe the various types of floors used in a cellular double bottom, and state where they are used.

31. Show how an aluminium superstructure is fastened to a steel deck. Explain why special precautions must be taken and what would happen if no such measures were taken.

32. Sketch and describe briefly: (a) bilge keel, (b) duct keel, (c) chain locker, (d) hawse pipe.

33. Define *hogging* and *sagging*. What members of the vessel are affected by these conditions? State the stresses in these members in each condition.

*34. With reference to a bulk-ore carrier explain how: (a) the ship is constructed to resist concentrated loads, (b) the GM is maintained at an acceptable value.

35. Define the following: gross tonnage; net tonnage; propelling power allowance.

36. Sketch and describe a sternframe. What material is used in its construction and why is this material suitable?

37. Describe with the aid of sketches the arrangements to withstand pounding in a ship.

38. Sketch any two of the following, giving approximate sizes: (a) a peak tank top manhole, (b) a windlass bed, (c) a bilge keel for a large, ocean going liner, (d) a bollard.

*39. Explain the essential constructional differences between the following types of vessels: (a) container ship, (b) ore carrier, (c) bulk oil carrier.

40. Describe the methods adopted in large passenger vessels to prevent the spread of fire. Show how this is accomplished in way of stairways and lift trunks.

*41. Describe with the aid of diagrammatic sketches TWO of the following systems used for transporting liquefied gas in bulk: (a) free-standing prismatic tanks, (b) membrane tanks, (c) free-standing spherical tanks.

42. A ship suffers stern damage due to collision with a quay. How would the ship be inspected to determine the extent of the damage? If the propeller were damaged state the procedure in fitting the spare propeller.

43. What important factors are involved before new tonnage can be called a *classified* ship?

44. Sketch and describe two types of modern rudder. How are they supported in the ship?

*45. 'In water survey' of large ships is now accepted under certain conditions as an alternative to dry docking. Discuss the main items to be inspected on the ship's hull during survey while the vessel is afloat.
   Briefly describe a remote controlled survey system which could be utilised for the examination of the ship's flat bottom.

46. Describe the destructive and non-destructive tests which may be carried out on welding material or welded joints.

47. Why may tank cleaning be dangerous? State any precautions which should be taken. How does the density of the gas vary throughout the tank during cleaning?

48. A vessel has taken a sudden list in a calm sea. What investigations should be made to ascertain the cause and what steps should be taken to right the vessel?

*49. Define with reasons the main purpose of each of the following practices:

(i) Use of neoprene washers in the connection between aluminium superstructures and ships' main structure.

(4 marks)

(ii) Attachment of anodic blocks to the underwater surface of a hull.   (3 marks)

(iii) External shotblasting and priming of hull plating.

(3 marks)

*50. Underwater hull survey of large ships is now permitted on occasions as an alternative to dry docking.

(i) State with reason what parts in particular should be inspected during such a survey.   (5 marks)

(ii) Describe briefly how such a survey is conducted from a position on board the vessel concerned.   (5 marks)

*51. (i) Sketch in diagrammatic form a transverse thrust unit which derives its thrust from a variable direction water jet.   (4 marks)

(ii) Explain how the unit operates.   (4 marks)

(iii) Compare the advantages of such units with those in which the thrust is produced by propeller(s) in a cross tunnel.   (2 marks)

*52. (i) Make a diagrammatic sketch of a watertight door, frame and closing gear, showing the manner of attachment to the bulkhead and the additional reinforcement carried by the bulkhead to compensate for the aperture.   (6 marks)

(ii) Explain how watertightness of the door/frame mating surfaces is ensured when the door is closed with a hydrostatic pressure tending to force the faces apart.   (2 marks)

(iii) State how upon failure of the primary means of closure, the door can be closed.   (2 marks)

*53. With reference to hull resistance evaluate the contribution made to its reduction by the following practices:
  (i) Abrasive blasting of hull plating, initially before paint application and during service.   (3 marks)
  (ii) Self polishing underwater copolymer antifouling coatings.   (3 marks)
  (iii) Impressed electrical current.   (2 marks)
  (iv) Biocide dosage.   (2 marks)

*54. Suggest with reasons whether 'build up' by welding, patching, cropping or plate replacement is best suited to the following defects:
  (i) Perforated hollow rudder.   (3 marks)
  (ii) Bilge keel partially torn away from hull.   (3 marks)
  (iii) Hull pierced, together with heavy indentation of bow below hawse pipe.   (4 marks)

*55. (i) Compare the advantages and disadvantages of single and twin screw propulsion purely from the point of view of stern and propeller arrangements alone.   (3 marks)
  (ii) Give reasons for the introduction of the ducted propeller.   (4 marks)
  (iii) Explain why blade tip damage on conventional propellers should be made good as soon as possible, when some ducted propellers operate satisfactorily with truncated blades.   (3 marks)

*56. With reference to stabiliser fins which either fold or retract into hull apertures:
  (i) Make a simplified sketch of the essential features of the activating gear for both fin extension and attitude.   (5 marks)
  (ii) Explain how it operates.   (5 marks)

*57. Suggest with reasons why the following conditions can contribute to reduction in ship speed:
  (i) Damaged propeller blades.   (2 marks)
  (ii) Indentation of hull plating.   (2 marks)
  (iii) Hole in hollow rudder plating.   (2 marks)
  (iv) Ship in ballast.   (2 marks)
  (v) Heavily fouled hull.   (2 marks)

INDEX

A

A-class bulkheads 133
Accommodation 12
Active controlled tanks 155
Active fin stabilisers 152
Advantages and disadvantages
   of welding 31
After end arrangement 82 *et seq.*
After peak 2, 77
Alternative tonnage 129
Aluminium alloy 39, 117, 166
   ,,   ,, sections 27
Anchors and cables 79
Angle bar 24
Anodes, sacrificial 165
Arc welding, argon 30
   ,,   ,, metallic 28
Arctic D steel 38
Argon arc welding 30
Assignment of class 136

B

B-class bulkheads 133
Balanced rudder 89
Ballast tanks 2, 41, 68
Beam knees 20, 48
Beams 19, 53
   ,, cant 82
   ,, longitudinal 53
   ,, strong 173
Bending moments 15, 17
   ,,   longitudinal 15 *et seq.*
   ,,   transverse 19
Bergensfjord, m.s. 166
Bilge blocks 21
   ,, keels 151
   ,, radius 14
   ,, shores 21
Boil-off 119
Bossings 93
Bottom longitudinals 101, 112
   ,, shell 50
Bow height 121
Bow thrusters 146
Bracket floor 44
Breadth extreme 13
   ,,   moulded 13
   ,,   tonnage 127
Brine trap 161
Brittle fracture 39
Built pillars 54
Bulb angle 25
   ,, plate 25
Bulbous bow 78
Bulk carriers 5, 110 *et seq.*

Bulkheads 61 *et seq.*
   ,, A and B-class 133
   ,, centreline 69
   ,, collision 75
   ,, corrugated 71, 103
   ,, deep tank 69
   ,, longitudinal 4, 101, 110
   ,, non-watertight 71
   ,, oiltight 103
   ,, swedged 72
   ,, transverse 61, 103
   ,, watertight 61
Bulwark 51
Buoyancy curve 15
Butt strap 137
   ,, welding 30

C

Cable gland 64
   ,, lifter 80
   ,, stopper 80
Camber 13
Cant beam 82
   ,, frame 82
Cargo hatch 3, 55
   ,, liner 2
   ,, piping 105
   ,, pumps 105
   ,, tramp 4
Casings, engine 171 *et seq.*
Cathodic protection 165
Ceiling 42
Centre girder 42
Centreline bulkhead 69
Chain locker 2, 79
   ,, pipes 79
Channel bar 26
Chemical carriers 10
Classification of ships 135 *et seq.*
   ,,   ,, societies 135
Coamings, hatch 55
Cofferdams 4, 47
Coffin plate 85
Colliers 6
Collision bulkhead 75
Conditions of assignment 125
Container ships 8, 121, *et seq.*
Continuity of structure 104, 138
Contra-rotating propellers 144
Controllable pitch propellers 143
Corrosion 162, *et seq.*
   ,, bi-metallic 165
   ,, prevention of 164
Corrugated bulkheads 71, 103
Covers, hatch 4, 55
Crack arrestors 137

Crane 173
,, rail support 173
Crew spaces 12
Crude oil washing 109
Cruiser stern 82 *et seq.*

D
Davits 131
Deadweight 14
Deck 52 *et seq.*
,, beams 53
,, camber 13
,, girders 53
,, line 125
,, longitudinals 53
,, machinery 54
,, plating 52
,, sheathing 53
Deep tanks 4, 68
,, ,, for water ballast 69
,, ,, for oil 69
Depth extreme 13
,, moulded 13
,, tonnage 127
Design and maintenance 167
,, of welded structure 35
Destructive testing 32
De-wedging device 65
Dimensions 12
Discontinuities 138
Displacement 14
Docking brackets 21, 103
,, stresses 21
Doors, casing 171
,, hinged 68
,, watertight 65
Double bottom 41 *et seq.*
,, ,, in machinery
space 46
,, ,, internal
structure 42
Drainage of insulated spaces 161
Draught extreme 13
,, moulded 13
Drydock, examination in 169
,, ,, support of ship in 21
Duct keel 21, 45

E
Edge preparation 30
Electrodes 28
Electrolytic action 162
Emergency repairs 170
End connections of
longitudinals 100
Engine casing 171 *et seq.*
,, room crane 173
Equipment 136
Examination in dry dock 169

Examination Questions,
Class 1 181 *et seq.*
,, Questions,
Class 2 176 *et seq.*
Extra notch-tough steel 37, 99

F
Fabricated sternframe 85
Faults in welded joints 34
Fillet welds 30
Fin stabiliser 152
Fire protection 133 *et seq.*
Flame cleaning 164
Flat plate keel 18, 50
Floors, bracket 44
,, solid 43
,, watertight 42
Fore end construction 73 *et seq.*
,, peak 2, 75
Forecastle 4
Fouling 167 *et seq.*
Frames, main 47
,, side 19, 47
,, 'tween deck 48
,, web 49
Framing, combined 102
,, longitudinal 99
,, transverse 47
Freeboard 13, 123 *et seq.*
,, conditions of
assignment 125
,, markings 125
,, surveys 127
,, tabular 123
Freeing ports 52
Fresh water allowance 125
Fully pressurised tanks 113
Fully refrigerated tanks 115
Funnel 174

G
Gangway, longitudinal 4, 126
Garboard strake 50
Gas venting system 108
Girders centre 42
,, deck 53
,, horizontal 69
,, intercostal 43
,, side 42
Grades of steel 37
Gravity davit 131
Gross tonnage 127
Gudgeons 85

H
Hatches 55 *et seq.*
,, coamings 55
,, covers 56
,, deep tank 58
,, oiltight 104

Hatches steel   57
,,   watertight   58
,,   webs   56
Hawse pipes   79
Heating coils   108
Higher tensile steel   38
Hogging   16
Holding down bolts   46
Holds, cargo   2
Hopper tanks   5, 113
Hydraulic thrust unit   148

I
Impressed current system   166
Inerting   120
Inner bottom   41
Intercostal girders   43
Intermittent welding   30
Internal structure of double
   bottom   42
Insulants   159
Insulated hatch plug   161
Insulation   117, 159 et seq.
,,   fire   133
In-water survey   138

J
Joint, types of   30
Joists   26

K
Keel   18, 50
,,   bilge   151
,,   blocks   21
,,   duct   21, 45
Kort nozzle   95
,,   ,,   rudder   97

L
Lap joint   30
Length between perpendiculars   13
,,   overall   13
,,   tonnage   127
Lifesaving
   appliances   131 et seq.
Lifeboats   131
Liferafts   131
Lightening holes   42
Lightweight   14
Liquefied gas carrier   9, 113 et seq.
Lloyd's Register of
   Shipping   135 et seq.
Load diagram   16
Load Line Rules   123
Longitudinal beams   53
,,   bending   15 et seq.
,,   bulkhead   4, 101, 110
,,   frames   44, 99
,,   gangway   4, 126
,,   material   18

Longitudinal strength   15

M
Machinery casings   171
,,   space   2
,,   space tonnage   128
Magnetic crack detection   34
Main vertical zones   134
Maintenance   167
Manhole cover   42
Margin plate   42
Material, longitudinal   18
,,   non-combustible   133
,,   transverse   19
Materials: 37 et seq.
   Aluminium alloys   38
   Arctic D steel   38
   Higher tensile steel   38
   Mild steel   37
Membrane tanks   117
Metallic arc welding   28
Midship section, bulk
   carrier   6, 112
Midship section cargo vessel   60
,,   ,,   chemical carrier   10
,,   ,,   collier   8
,,   ,,   container ship   121
,,   ,,   cylindrical
      tank system   114
,,   ,,   liquefied gas
      carrier   11
,,   ,,   oil tanker   5, 100, 101
,,   ,,   ore carrier   6, 111
,,   ,,   prismatic tank
      system   116
,,   ,,   showing
      dimensions   12
,,   ,,   spherical tank
      system   115
Mild steel   37
Mill scale, removal of   164
Modified tonnage   128

N
Net tonnage   127
Non-combustible material   133
Non-destructive testing   33
Non-watertight bulkheads   71
Notch-tough steel   37, 99
Nozzle, Kort   95

O
Oil fuel tanks   2, 41, 69
Oil tankers   4, 99 et seq.
,,   ,,   combined framing   102
,,   ,,   longitudinal framing   99
Oiltight bulkhead   103
,,   hatch   104
Open water stern   91
Ore carriers   5, 110

P
Painting 165
Panting 23
,, arrangements 73
Passenger ships 1
,,   ,, fire protection 134
Passive tanks 154
Peaks 2, 75
Pickling 164
Pillars 53
Pintles 87
Piping, main cargo 105
Plating, deck 18, 52
,, shell 18, 50
,, stem 73
Pounding 23
,, arrangements 72
Prismatic tanks 116
Profile, bulk carrier 7
,, cargo liner 3
,, cargo tramp 3
,, collier 7
,, container ship 7
,, liquefied gas
        carrier 11
,, oil tanker 3
,, roll-on/roll-off vessel 9
,, showing dimensions 11
Propeller, examination of 170
Propellers 141
,, contra-rotating 144
,, controllable pitch 143
,, vertical axis 145
Propelling Power Allowance 128
Protection, cathodic 165
,, fire 133 *et seq.*
Pump rooms 4, 105
Pumping arrangements of oil
    tankers 105
Pumps, cargo 105, 120
,, stripping 105

R
Racking 20
Radiographic testing of welds 33
Rails, crane 173
Raised quarter deck 8
Reduction of roll 151 *et seq.*
Refrigerated vessels 2
,,   ,, insulation of 159
Repair 170
Rise of floor 14
Rolled steel sections 24 *et seq.*
Rolling and stabilisation 150 *et seq.*
Roll-on/Roll-off vessels 9
Rotating cylinders 98
Round of beam—see camber 13
Rounded gunwale 18
Rudder carrier 88

Rudders 84 *et seq.*
,, balanced 89
,, Kort nozzle 97
,, semi-balanced 92
,, spade 91
,, unbalanced 85

S
Sacrificial anodes 165
Sagging 17
Scantlings 137
Scuppers 126
Seating, winch 54
Sections, aluminium 27
,, steel 24 *et seq.*
Self polishing copolymers 168
Self trimming collier 8
Semi-pressurised tanks 114
Settling tanks 2
Shaft stools 95
,, tunnel 2, 94
Sheathing, deck 53
Sheer 14
Sheerstrake 18, 50
Shell plating 50 *et seq.*
Ship types 1 *et seq.*
Shot blasting 165
Side framing 20, 47, 102
,, girders 18, 42
,, shell 18, 50
,, shores 22
Solid floors 43
Solid pillars 54
Spade rudder 91
Spectacle frame 93
Sprinkler system 134
Stays, funnel 174
Steel, extra notch-tough 37, 99
,, hatch covers 57
,, sections 24 *et seq.*
Stem bar 73
,, construction 73, 78
,, plating 73
Stern, cruiser 82
,, transom 84
Sternframe, cast 89
,, fabricated 85
,, open water 91
,, twin screws 92
Stiffeners, bulkhead 62
Still water bending 15
Strength, longitudinal 15
,, transverse 19
Stresses 15 *et seq.*
Stringer plate 18, 50
Stripping pumps 105
Strong beams 173
Superstructure, aluminium 166
Surface preparation 164

Surveys, annual 127, 136
  ,,   in-water 138
  ,,   special 136
Swedged bulkheads 72

T
T bars 27
Tabular freeload 121
Tailshaft weardown 138, 170
Tail flaps 98
Tank margin 42
  ,, side brackets 20, 47
Tank stabiliser 154
  ,, top 2, 41
Tankers—see oil tankers 99 *et seq.*
Tanks, deep 4, 68
  ,,   double bottom 2, 41
  ,,   peak 2, 75
  ,,   testing of 42
Terms in general use 12 *et seq.*
Testing, destructive 32
  ,,   non-destructive 33
  ,,   of tanks 42
  ,,   of welds 32
Tonnage 127 *et seq.*
  ,,   1967 Rules 127
  ,,   1982 Rules 129
  ,,   alternative 129
  ,,   deductions 127
  ,,   enclosed spaces 130
  ,,   gross 127
  ,,   mark 129
  ,,   modified 128
  ,,   net 127
  ,,   underdeck 127
Torsion boxes 122, 140
Transom floor 85
  ,,   stern 84
Transverse bending 19
  ,,   bulkheads 61, 103
  ,,   framing 47

Transverse material 19
  ,,   strength 19
  ,,   webs 101, 112
Tumble home 14
Tunnel, shaft 2, 94
Tween decks 2
Types of ships 1 *et seq.*
Type-A ship 121
Type-B ship 122

U
Unbalanced rudder 85
Ultrasonic testing of welds 34
Underdeck tonnage 127
Uptakes, funnel 174

V
Vertical axis propellers 145
Vibration 156 *et seq.*

W
Water ballast tanks 2, 41, 69
Water pressure 19, 62
Watertight bulkheads 61 *et seq.*
  ,,   doors 65 *et seq.*
Wave bending 18
Weathering 164
Web frames 49
Weight curve 15
Welded joints 30
  ,,   plate collars 40
Welding, advantages and dis-
         advantages of 31
Welding, argon arc 30
  ,,   faults 34
  ,,   metallic arc 28
  ,,   testing 32
Wood sparring 49

X
X-rays 33

REED'S MARINE ENGINEERING SERIES

Vol 1 Mathematics
Vol 2 Applied Mechanics
Vol 3 Applied Heat
Vol 4 Naval Architecture
Vol 5 Ship Construction
Vol 6 Basic Electrotechnology
Vol 7 Advanced Electrotechnology
Vol 8 General Engineering Knowledge
Vol 9 Steam Engineering Knowledge
Vol 10 Instrumentation and Control Systems
Vol 11 Engineering Drawing
Vol 12 Motor Engineering Knowledge

Reeds Instruments and Control Systems for Deck Officers
Reeds Mathematical Tables and Engineering Formulae
Reeds Marine Distance Tables
Reeds Sextant Simplified
Reeds Skipper's Handbook
Reeds Maritime Meteorology
Reeds Marine Insurance
Reeds Marine Surveying
Reeds Sea Transport – Operation and Economics

These books are obtainable from nautical
booksellers, chandlers or direct from:

Macmillan Distribution Ltd
Brunel Road
Houndsmill
Basingstoke
RG21 6XS

Tel: 01256 302699
Fax: 01256 812521/812558
Email: direct@macmillan.co.uk